Electronic Devices and Circuits

AUTHOR

Prof. Ankesh Jain
Assistant Professor,
Department of Electrical Engineering
Indian Institute of Technology, Delhi

REVIEWER

Prof. Shivam Verma
Associate Professor,
Department of Electronics and Communication Engineering
Indian Institute of Technology, Varanasi

KHANNA BOOK PUBLISHING CO. (P) Ltd.
4C/4344, Ansari Road, Darya Ganj,
New Delhi, India

All India Council for Technical Education
Nelson Mandela Marg, Vasant Kunj,
New Delhi, India

Dear Readers,

To prevent the piracy, this book is secured with HIGH SECURITY HOLOGRAM on the front title cover. In case you don't find the hologram on the front cover title, please write us to at contact@khannabooks.com or whatsapp us at +91-99109 09320 and avail special gift voucher for yourself.

Specimen of Hologram on front Cover title:

Moreover, there is a SPECIAL DISCOUNT COUPON for you with EVERY HOLOGRAM.

How to avail this SPECIAL DISCOUNT:

Step 1: Scratch the hologram

Step 2: Under the scratch area, your "coupon code" is available

Step 3: Logon to www.khannabooks.com

Step 4: Use your "coupon code" in the shopping cart and get your copy at a special discount

Step 5: Enjoy your reading!

ISBN: 978-93-5538-720-2
Book Code: DIP454EN

Electronic Devices and Circuits
by Prof. Ankesh Jain

First Edition: 2025

Originally Published by:
All India Council for Technical Education, Ministry of Education, GOI

Exclusively Published & Distributed by:
Khanna Book Publishing Co. (P) Ltd.
4C/4344, Ansari Road, Daryaganj,
New Delhi (INDIA)

Visit us at: **www.khannabooks.com**
Write us at: **contact@khannabooks.com**

To view complete list of books, Please scan the QR Code: KPH

Printed in Bharat.

प्रो. टी. जी. सीताराम
अध्यक्ष
Prof. T. G. Sitharam
Chairman

सत्यमेव जयते

आज़ादी का अमृत महोत्सव

अखिल भारतीय तकनीकी शिक्षा परिषद्
(भारत सरकार का एक सांविधिक निकाय)
(शिक्षा मंत्रालय, भारत सरकार)
नेल्सन मंडेला मार्ग, वसंत कुंज, नई दिल्ली–110070
दूरभाष : 011–26131498
ई–मेल : chairman@aicte-india.org

ALL INDIA COUNCIL FOR TECHNICAL EDUCATION
(A STATUTORY BODY OF THE GOVT. OF INDIA)
(Ministry of Education, Govt. of India)
Nelson Mandela Marg, Vasant Kunj, New Delhi-110070
Phone : 011-26131498
E-mail : chairman@aicte-india.org

FOREWORD

Engineers are the backbone of any modern society. They are the ones responsible for the marvels as well as the improved quality of life across the world. Engineers have driven humanity towards greater heights in a more evolved and unprecedented manner.

The All India Council for Technical Education (AICTE), have spared no efforts towards the strengthening of the technical education in the country. AICTE is always committed towards promoting quality Technical Education to make India a modern developed nation emphasizing on the overall welfare of mankind.

An array of initiatives has been taken by AICTE in last decade which have been accelerated now by the National Education Policy (NEP) 2020. The implementation of NEP under the visionary leadership of Hon'ble Prime Minister of India envisages the provision for education in regional languages to all, thereby ensuring that every graduate becomes competent enough and is in a position to contribute towards the national growth and development through innovation & entrepreneurship.

One of the spheres where AICTE had been relentlessly working since past couple of years is providing high quality original technical contents at Under Graduate & Diploma level prepared and translated by eminent educators in various Indian languages to its aspirants. For students pursuing 2nd year of their Engineering education, AICTE has identified 87 books, which shall be translated into 12 Indian languages - Hindi, Tamil, Gujarati, Odia, Bengali, Kannada, Urdu, Punjabi, Telugu, Marathi, Assamese & Malayalam. In addition to the English medium, books in different Indian Languages are going to support the students to understand the concepts in their respective mother tongue.

On behalf of AICTE, I express sincere gratitude to all distinguished authors, reviewers and translators from the renowned institutions of high repute for their admirable contribution in a record span of time.

AICTE is confident that these outcomes based original contents shall help aspirants to master the subject with comprehension and greater ease.

(Prof. T. G. Sitharam)

ACKNOWLEDGEMENT

The author is grateful to the authorities of AICTE, particularly Prof. T. G. Sitharam, Chairman; Dr. Abhay Jere, Vice-Chairman; Prof. Rajive Kumar, Member-Secretary; and Dr. Ramesh Unnikrishnan, Advisor-II and Dr. Sunil Luthra, Director, Training and Learning Bureau. Their collective vision and planning were instrumental in the publication of the book on *Electronic Devices and Circuits*. I sincerely acknowledge the valuable contributions of the reviewer of the book Prof. Shivam Verma, Assistant Professor, Department of Electronics and Communication Engineering, Indian Institute of Technology, Varanasi for making it students' friendly and giving a better shape in an artistic manner.

I am indebted to all my teachers and professors who have taught me and directly or indirectly this book has inherited some of that learning. I would like to especially thank my PhD supervisor, Prof. Shanthi Pavan who has always inspired me to aim high. I am also indebted to my parents (JadugarJinesh Jain and Indra Jain) who have also inspired me and encouraged me to follow the path of learning. I would like to dedicate this book to my wife (Tripti Jain) who has always inspired me, encouraged me and allowed me to work on this book while taking away the time from her. Apart from this I am also thankful to all my friends, nears and dears who have always helped me in coming so far in the life.

This book is an outcome of various suggestions of AICTE members, experts and authors who shared their opinion and thought to further develop the engineering education in our country. Acknowledgements are due to the contributors and different workers in this field whose published books, review articles, papers, photographs, footnotes, references and other valuable information enriched us at the time of writing the book.

Ankesh Jain

PREFACE

The book titled "Electronic devices and circuits" is an outcome of the rich experience of our teaching of circuit theory, semiconductor devices and design related courses. The initiation of writing this book is to expose fundamental of electronic devices and circuits to the engineering students, the foundation developed in this book will enable them to get an insight of the subject. Keeping in mind the purpose of wide coverage as well as to provide essential supplementary information, we have included the topics recommended by AICTE, in a very systematic and orderly manner throughout the book. Efforts have been made to explain the fundamental concepts of the subject in the simplest possible way.

During the process of preparation of the manuscript, I have considered the various standard text books and accordingly I have developed sections like critical questions, solved and supplementary problems etc. While preparing the different sections emphasis has also been laid on definitions and laws and also on comprehensive synopsis of formulae for a quick revision of the basic principles. The book covers all types of medium and advanced level problems and these have been presented in a very logical and systematic manner. The gradations of those problems have been tested over many years of teaching to a wide variety of students.

Apart from illustrations and examples as required, I have enriched the book with numerous solved problems in every unit for proper understanding of the related topics. It is important to note that in this book, I have included the relevant laboratory practical. In addition, besides some essential information for the users under the heading "Know More" I have clarified some essential basic information in the appendix and annexure section.

As far as the present book is concerned, "Electronic devices and circuits" is meant to provide a thorough grounding in semiconductor design on the topics covered. This part of the book will prepare engineering students to apply the knowledge of electronics to tackle 21st century and onward engineering challenges and address the related aroused questions. The subject matters are presented in a constructive manner so that an Engineering degree prepares students to work in different sectors or in national laboratories at the very forefront of technology.

We sincerely hope that the book will inspire the students to learn and discuss the ideas behind basics of electronic devices and circuits and will surely contribute to the development of a solid foundation of the subject. We would be thankful to all beneficial comments and suggestions which will contribute to the improvement of the future editions of the book. It gives me immense pleasure to place this book in the hands of the teachers and students. It was indeed a big pleasure to work on different aspects covering in the book.

Ankesh Jain

OUTCOME BASED EDUCATION

For the implementation of an outcome based education the first requirement is to develop an outcome based curriculum and incorporate an outcome based assessment in the education system. By going through outcome based assessments, evaluators will be able to evaluate whether the students have achieved the outlined standard, specific and measurable outcomes. With the proper incorporation of outcome based education there will be a definite commitment to achieve a minimum standard for all learners without giving up at any level. At the end of the programme running with the aid of outcome based education, a student will be able to arrive at the following outcomes:

Programme Outcomes (POs) are statements that describe what students are expected to know and be able to do upon graduating from the program. These relate to the skills, knowledge, analytical ability attitude and behaviour that students acquire through the program. The POs essentially indicate what the students can do from subject-wise knowledge acquired by them during the program. As such, POs define the professional profile of an engineering diploma graduate.

National Board of Accreditation (NBA) has defined the following seven POs for an Engineering diploma graduate:

PO-1: Basic and Discipline specific knowledge: Apply knowledge of basic mathematics, science and engineering fundamentals and engineering specialization to solve the engineering problems.

PO-2: Problem analysis: Identify and analyses well-defined engineering problems using codified standard methods.

PO-3: Design/development of solutions: Design solutions for well-defined technical problems and assist with the design of systems components or processes to meet specified needs.

PO-4: Engineering Tools, Experimentation and Testing: Apply modern engineering tools and appropriate technique to conduct standard tests and measurements.

PO-5: Engineering practices for society, sustainability and environment: Apply appropriate technology in context of society, sustainability, environment and ethical practices.

PO-6: Project Management: Use engineering management principles individually, as a team member or a leader to manage projects and effectively communicate about well-defined engineering activities.

PO-7: Life-long Learning: Ability to analyse individual needs and engage in updating in the context of technological changes.

COURSE OUTCOMES (CO)

After completion of the course the students will be able to:

CO-1: Know about basic operations of semiconductor devices and its involved semiconductor physics

CO-2: Basic analysis of circuits designed using semiconductor devices

CO-3: Basic understanding of circuit design using semiconductor devices

CO-4: Basic understanding of design, simulation and experiment flow in electronic design

CO-5: To apply knowledge of electronic design and analysis in real life problems

Course Outcomes	EXPECTED MAPPING WITH PROGRAMME OUTCOMES (1- Weak Correlation; 2- Medium correlation; 3- Strong Correlation)											
	PO-1	PO-2	PO-3	PO-4	PO-5	PO-6	PO-7	PO-8	PO-9	PO-10	PO-11	PO-12
CO-1	3	2	2	1	1	1	-	-	-	-	-	3
CO-2	3	3	2	1	1	1	-	-	-	-	-	3
CO-3	3	3	3	3	3	2	-	-	1	2	2	1
CO-4	3	3	3	3	3	3	-	-	1	2	-	3
CO-5	3	3	3	3	3	3	-	-	2	2	2	3

GUIDELINES FOR TEACHERS

To implement Outcome Based Education (OBE) knowledge level and skill set of the students should be enhanced. Teachers should take a major responsibility for the proper implementation of OBE. Some of the responsibilities (not limited to) for the teachers in OBE system may be as follows:

- Within reasonable constraint, they should manoeuvre time to the best advantage of all students.
- They should assess the students only upon certain defined criterion without considering any other potential ineligibility to discriminate them.
- They should try to grow the learning abilities of the students to a certain level before they leave the institute.
- They should try to ensure that all the students are equipped with the quality knowledge as well as competence after they finish their education.
- They should always encourage the students to develop their ultimate performance capabilities.
- They should facilitate and encourage group work and team work to consolidate newer approach.
- They should follow Blooms taxonomy in every part of the assessment.

Bloom's Taxonomy

Level	Teacher should Check	Student should be able to	Possible Mode of Assessment
Creating	Students ability to create	Design or Create	Mini project
Evaluating	Students ability to Justify	Argue or Defend	Assignment
Analysing	Students ability to distinguish	Differentiate or Distinguish	Project/Lab Methodology
Applying	Students ability to use information	Operate or Demonstrate	Technical Presentation/ Demonstration
Understanding	Students ability to explain the ideas	Explain or Classify	Presentation/Seminar
Remembering	Students ability to recall (or remember)	Define or Recall	Quiz

GUIDELINES FOR STUDENTS

Students should take equal responsibility for implementing the Outcome Based Education (OBE). Some of the responsibilities (not limited to) for the students in OBE system are as follows:

- Students should be well aware of each Unit Outcome (UO) before the start of a unit in each and every course.
- Students should be well aware of each Course Outcome (CO) before the start of the course.
- Students should be well aware of each Program Outcome (PO) before the start of the program.
- Students should think critically and reasonably with proper reflection and action.
- Learning of the students should be connected and integrated with practical and real life consequences.
- Students should be well aware of their competency at every level of OBE.

ABBREVIATIONS AND SYMBOLS

List of Abbreviations

General Terms			
Abbreviations	**Full form**	**Abbreviations**	**Full form**
LED	Light Emitting Diode	MOS	Metal oxide semiconductor
AC	Alternating current	NMOS	n-type MOS
DC	Direct current	PMOS	p-type MOS
PIV	Peak Inverse Voltage	CS	Common Source
BJT	Bipolar Junction Transistor	CD	Common Drain
FET	Field Effect Transistor	CG	Common Gate
KVL	Kirchhoff Voltage Law	UJT	Unijunction Transistor
KCL	Kirchhoff Current Law	SCR	Silicon Controlled Rectifier
CE	Common Emitter	IGBT	Insulated Gate Bipolar Transistor
CB	Common Base	DIAC	Diode for alternating current
CC	Common Collector	TRIAC	Triode for alternating current
h parameter	Hybrid parameter	SNR	Signal to Noise Ratio
JFET	Junction Field Effect Transistor	THD	Total Harmonic Distortion
IGFET	Insulated Gate Field Effect Transistor	SFDR	Spurious free dynamic range
		LHP	Left half plane
MOSFET	Metal Oxide Semiconductor Field Effect Transistor	RHP	Right half plane
		CMOS	Complementary Metal oxide semiconductor
IC	Integrated Circuit		

List of Units and Symbols

Symbols	Description	Symbols	Description
Si	Silicon	K	Kelvin
Ge	Germanium	ϵ_s	Permittivity of silicon
P	Phosphorous	β_F	Common emitter current gain
B	Boron	α_F	Common base current gain
In	Indium	q	Magnitude of the Charge of an electron
Ga	Gallium	V_T	Volt equivalent of temperature
Al	Aluminum	g_m	Small signal
r_π	transconductance Small signal input impedance	V_t	Threshold voltage
r_o	Small signal output impedance	λ	Channel length modulation parameter
V_A	Early voltage	I_L	Latching current
μ_n	Mobility of electron	I_H	Holding current
μ_p	Mobility of holes		

LIST OF FIGURES

CONTENTS

1 Semiconductor Diodes

UNIT SPECIFICS

Through this unit we have discussed the following aspects:

- Introduction to semiconductors and types of semiconductors;
- P-N junction, Depletion and Biasing of device;
- Diodes, characteristic of diode, types of diode;
- Application of diodes
- Filter circuits used in rectification

The practical applications of the topics are discussed for generating further curiosity and creativity as well as improving problem solving capacity.

Besides giving a large number of multiple choice questions as well as questions of short and long answer types marked in two categories following lower and higher order of Bloom's taxonomy, assignments through a number of numerical problems, a list of references and suggested readings are given in the unit so that one can go through them for practice. It is important to note that for getting more information on various topics of interest some QR codes have been provided in different sections which can be scanned for relevant supportive knowledge.

After the related practical, based on the content, there is a "Know More" section. This section has been carefully designed so that the supplementary information provided in this part becomes beneficial for the users of the book. This section mainly highlights the initial activity, examples of some interesting facts, analogy, history of the development of the subject focusing the salient observations and finding, timelines starting from the development of the concerned topics up to the recent time, applications of the subject matter for our day-to-day real life or/and industrial applications on variety of aspects, case study related to environmental, sustainability, social and ethical issues whichever applicable, and finally inquisitiveness and curiosity topics of the unit.

RATIONALE

This chapter provides a basic introduction to semiconductor, its behaviour, its differences with other material. It also builds a foundation of semiconductor physics where operation and behaviour of a p-n junction is described in detail. The chapter provides a brief overview of different types of diode and discusses about several important applications of diode in detail.

PRE-REQUISITES

Mathematics: Algebra (Class XII)

Physics: Semiconductors (Class XII)

UNIT OUTCOMES

List of outcomes of this unit is as follows:

U1-O1: Basic understanding of semiconductors

U1-O2: Basic understanding of semiconductor operation and p-n junction

U1-O3: Diode, characterization of diodes, categories of diode

U1-O4: Application of diode

U1-O5: Experimental understanding of diode

Unit-1 Outcome	EXPECTED MAPPING WITH COURSE OUTCOMES (1 - Weak Correlation; 2 - Medium Correlation; 3 - Strong Correlation)				
	CO-1	CO-2	CO-3	CO-4	CO-5
U1-O1	3	1	1	1	3
U1-O2	3	1	1	1	1
U1-O3	3	2	3	2	2
U1-O4	1	3	3	3	3
U1-O5	1	3	3	3	3

1.1 INTRODUCTION

One can find electronic devices embedded in almost everything around us signifying the importance of electronic systems. In fact it is not wrong to say that electronics has become an integral part of today's life. Hence, it is useful to know more about the basic functioning and operation of these devices. The operation of electronic devices is based on flow of electrons. An electron is a negatively charged particle which is a part of atom and revolves around its nucleus. The electrons in outermost orbit of an atom which is also called as valence shell, take part in bonding and are shared with other atoms through covalent or ionic bands. These bonds keep electrons tightly associated with these atoms. However, electrons can come out of valence shell after gaining sufficient kinetic energy and thus can become free. A free electron is no longer associated with a particular atom and can flow freely in an electric field. This way free electron can results in flow of charge carrier and thus can constitute an electric current.

1.1.1 Energy State, Band, Valence Band, Conduction Band and Bandgap

The energy state of electron in an atom is specific however when several atoms are brought into proximity then energy state of electron splits due to Pauli's exclusion principle and it forms a band of energy states. In a material topmost nearly filled band is defined as valence band while bottommost nearly empty band is defined as conduction band. The energy gap between valence band and conduction band is defined as bandgap.

1.1.2 Conductors, Insulators and Semiconductors

The materials can be classified based on availability of free electrons. Metals are materials with an abundant amount of free electrons and thus conduct easily and can have a large amplitude of current even in a small electric field. This is why metals are also known as conductors.

While on the other hand, there are materials which have very few free electrons or no free electrons at all and thus does not conduct well and are termed as insulators.

A set of materials which has its conductivity between metal and insulators are termed as semiconductors. Silicon (Si) and Germanium (Ge) are two example of semiconductor materials. The conductivity of semiconductor material increases with temperature which is in contrast to metals where conductivity decreases with temperature.

The difference between metal, insulator and semiconductor can also be explained through bandgap. The bandgap is energy difference between top of valence band and bottom of conduction band and hence it is essentially energy required by electron to move from valence band to conduction band where it becomes free to move. The metals have no bandgap as valence band and conduction band overlaps in it. Hence, it has plenty of free electrons. Insulators have significantly high bandgap and hence it has no or very few electrons in its conduction band and thus it has very few or no free electrons.

1.2. INTRINSIC AND EXTRINSIC SEMICONDUCTOR

Semiconductor materials are classified in following two categories,

1.2.1 Intrinsic Semiconductor

A pure semiconductor material such as Si or Ge is known as intrinsic semiconductor. Both Si and Ge has 4 electrons in its valence shell and each of these electron is shared with neighbouring Si (or Ge) atoms through a covalent bond forming a tetravalent crystalline Si (or Ge) structure as shown in Fig.1.1 and thus achieving inert structure in its outermost shell. Hence at 0 K there is no free electron available in an intrinsic semiconductor and it behaves like an insulator. However, if the temperature is increased then some of the electrons becomes free by achieving sufficient kinetic energy.

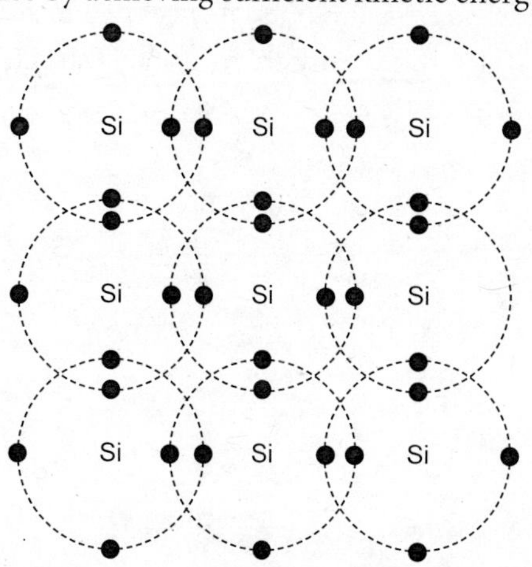

Fig.1.1: Two dimensional representation of tetravalent structure of silicon

The number of free charge carriers at room temperature can be increased in a semiconductor material by adding impurities from III or V group materials in the periodic table. This process of adding impurities is called as doping and the atoms belonging to III or V groups are called as dopant.

1.2.2 Extrinsic semiconductor

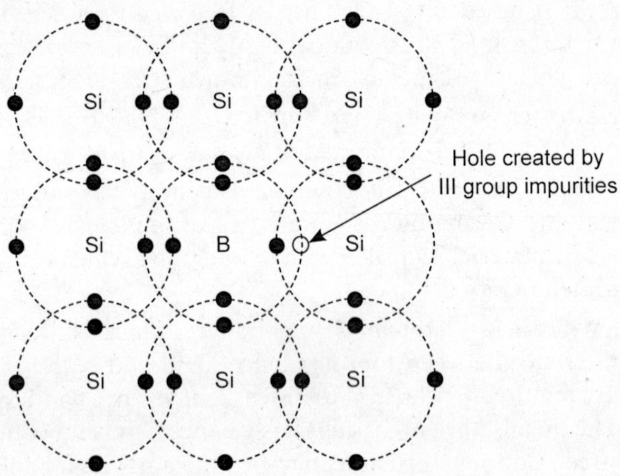

Fig.1.2: p-type semiconductor

The semiconductor material which has been added with specific impurities in it to alter its electrical properties are known as extrinsic semiconductors.

The addition of III group impurities such as Boron(B), Indium (In), Gallium (Ga) or Aluminium (Al) in an intrinsic semiconductor creates a vacancy in tetravalent crystalline structure of Si (or Ge) as dopant atom has only three electrons in its valence shell and thus can share its electrons only with three neighbouring Si (or Ge) atoms. This vacancy of electrons is defined as holes and has an effective positive charge. When an electron in valence shell occupy this vacancy and thereby creates an another vacancy, holes are effectively moved which constituent an electric current. As holes are positively charged so this type of semiconductor are termed as p-type semiconductor. The III group impurities are also called as acceptor dopant as they have one less electron in its valence shell to achieve a stable structure according to octet rule and by accepting an electron it can achieve a stable structure.

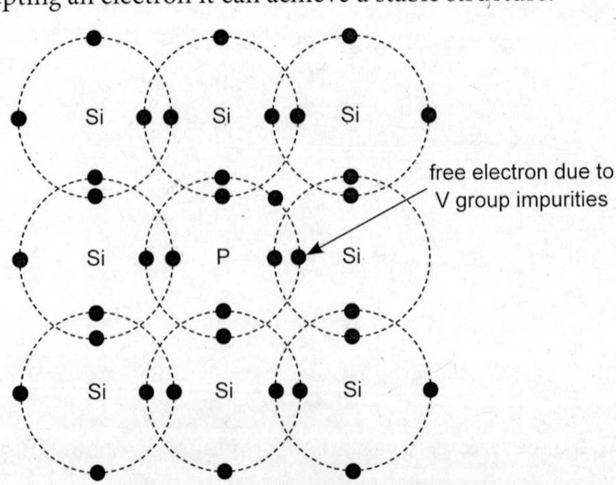

Fig.1.3: n-type semiconductor

If impurities from V group such as Phosphorous(P) is introduced in an intrinsic semiconductor material such as Silicon (Si) as shown in Fig.1.3 then four out of five electrons in valence shell of dopant atom (P) can form covalent bond with four neighbouring Si atoms however one electron remain weakly bounded with dopant atoms. This weakly bounded electrons can become free with much lesser thermal energy. Thus addition of V group impurities results in a large number of available free electrons. As the electrons are negatively charged, this type of semiconductor is termed as n-type semiconductor. The V group impurities are also called as donor dopant as they have one more electron in its valence shell than required to achieve a stable structure according to octet rule and by donating an electron it can achieve a stable structure.

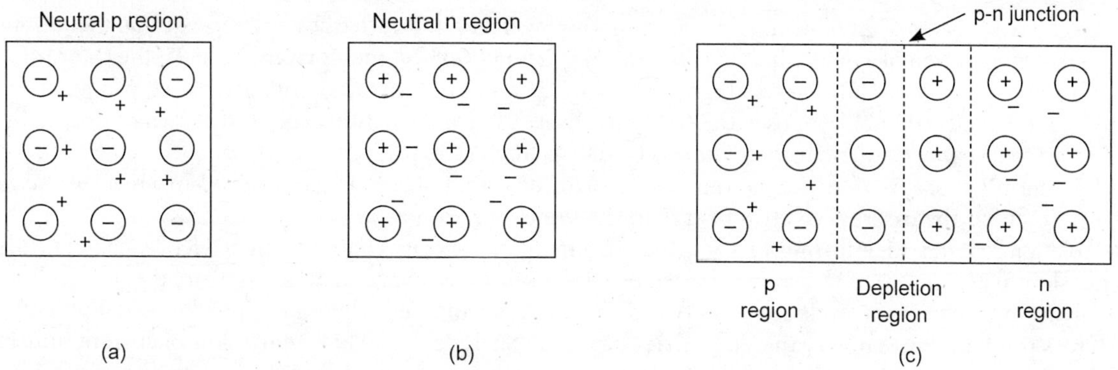

Fig.1.4: (a) p-type semiconductor (b) n-type semiconductor (c) p-n junction formed by bringing p-type and n-type semiconductors into contact of each other

Fig.1.4(a) and Fig.1.4(b) represent p-type semiconductor and n-type semiconductor respectively. Although each of these regions have a large number of free charge carriers however overall region is charge neutral as nucleus has equal and opposite charge.

1.3 P-N JUNCTION

A p-n junction is formed by bringing a p-type semiconductor material in contact with an n-type semiconductor material. When a p-type semiconductor material is brought into contact of an n-type semiconductor material then electrons which are majority charge carriers in a n-type semiconductor starts moving to p-type semiconductor and holes which are majority charge carriers in a p-type semiconductor start moving to n-type semiconductor due to diffusion. The electrons coming from n-type semiconductor combines with holes present in p-type semiconductor leaving the region near p-n junction as negatively charged and devoid of any mobile charge carriers. In similar way, the holes from p-type semiconductor move to n-type semiconductor and combine with electrons present in n-type semiconductor leaving the region near p-n interface in n-type semiconductor as positively charged and devoid of mobile charge carriers. This region at p-n interface which is devoid of any mobile charge carrier is termed as depletion region as it is depleted of the free charge carriers. This oppositely charged depletion region results in an electric field which opposes the movement of any further majority charge carrier from any side due to diffusion. The minimum potential needed to overcome this electric field to ensure the current flow through p-n junction is known as barrier potential or built in potential. An external potential source is required to overcome this barrier potential.

The width of the depletion region W under zero bias is given by (1.1)

$$W = \sqrt{\frac{2\epsilon_s}{q}V_o\left(\frac{1}{N_A}+\frac{1}{N_D}\right)} \tag{1.1}$$

Where ϵ_s is the permittivity of the semiconductor material, q is the charge of electron, V_o is the barrier potential, N_A is the concentration of acceptor dopant and N_D is the concentration of donor dopant. It is evident from this equation that higher doping results in a thinner depletion region.

1.4 BIASING

Biasing is a process of establishing set of currents and voltages in a device by applying external excitations to fix the operation of a device in a specific way. A p-n junction becomes forward biased if the p-terminal of a p-n junction is connected to a higher potential than its n-terminal. In forward biased p-n junction, the external electric field opposes the field established due to depletion region. If external potential is higher than barrier potential than current begins to flow. The p-n junction becomes reverse biased if the p-terminal of a p-n junction is connected to a lower potential than its n-terminal. In a reverse biased p-n junction, the external electric field is in the same direction as the electric field established due to depletion region which further restricts the movement of majority charge carriers. Hence, current does not flow to a large extent in a reverse biased p-n junction. Thus, the current in a p-n junction diode flows only from p-type semiconductor to n-type semiconductor that too when the applied external potential across p-n junction is higher than its barrier potential. Such devices which allow flow of current only in one direction are termed as unilateral devices. A p-n junction is a two terminal unilateral device which is categorised as diode.

1.5 DIODE

A Diode is a two terminal unilateral device which allows flow of current only in one direction. Its terminals are named as anode (positive terminal) and cathode (negative terminal). A diode can be thought of as a valve which allows flow only in one direction.

A diode is in forward bias if the potential difference between anode and cathode is positive. A forward biased diode act like a closed switch which conducts. A diode is in reverse bias if the potential difference between anode and cathode is negative. A reverse biased diode act like an open switch which does not conduct.

1.5.1 I-V Characteristic of the Diode

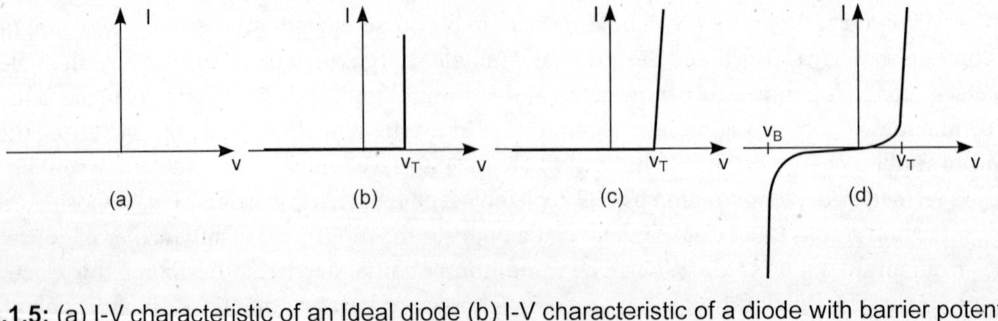

Fig.1.5: (a) I-V characteristic of an Ideal diode (b) I-V characteristic of a diode with barrier potential (c) I-V characteristic of diode including barrier potential and forward resistance (d) I-V characteristic of real diode

The I-V characteristic of an ideal diode is shown in Fig.1.5(a). It has zero current in reverse bias and zero voltage across it in forward bias. The I-V characteristic shown in Fig.1.5(b) models the effect of barrier potential which is also termed as cut-in voltage or threshold voltage of the diode. The characteristic can further be modified as the one shown in Fig.1.5(c) where finite slope of characteristic in forward bias region models the forward resistance of a diode.

The I-V characteristic of a p-n junction diode resembles with the characteristic shown in Fig.1.5(d). It has three distinct region. If the applied potential at anode is higher than cathode then the diode is in forward region and it conducts. If the applied potential at anode is smaller than the potential at cathode then the diode is in reverse bias. A reverse biased p-n junction diode restricts the flow of majority charge carriers however minority charge carriers can still flow across p-n junction as the overall field favours its flow. The minority charge carriers move through p-n junction by passing through depletion region where the electric field is generally high. This high electric field accelerates the electrons and holes passing through it. If reverse bias potential increases beyond a value V_B, the minority charge carriers gets enough energy to knock out electrons from valence band by breaking the covalent bond and generating electron-hole pairs. These electron-hole pair further boost the process by repeating it and thus flooding (avalanche) the p-n junction with large number of minority charge carrier resulting in large amount of current. This process is known as Avalanche breakdown and the voltage V_B is known as avalanche breakdown voltage. The avalanche breakdown is caused by impact ionization due of electron-hole pairs as discussed above. If excessive current generated during avalanche breakdown is not controlled through external method then diode may get permanently damaged due to overheating.

The behaviour of a semiconductor diode in constant current forward bias region and reverse bias region is given by Shockley diode equation given by (1.2)

$$I = I_S(e^{\frac{V_D}{nV_T}} - 1) \tag{1.2}$$

Where I_S is the reverse bias saturation current

V_D is the voltage across diode

V_T is the volt equivalent of temperature

n is the ideality factor which depends on type of semiconductor

Thus, diode is a nonlinear element as its I-V characteristic is nonlinear. However, small signal approximation of a nonlinear element can be linear which is much easier to analyze.

1.5.2 Small Signal Analysis

Let's consider a nonlinear relation which is given in (1.3)

$$y = f(x) \tag{1.3}$$

Where x is an independent variable and *y* is a dependent variable and *f* is the function which relates *x* to *y*. Now if value of *y* for a particular value of *x* say x_o is y_o, then to find the value of *y* for another value of x say $x_o + dx$ (1.3) has to be solved again which is a nonlinear equation and solving it is a relatively difficult or computationally intensive. However, if *dx* is very small then help of Taylor series expansion is taken to simplify this process. This expression is given in (1.4) which looks little more complicated.

$$f(x_o + dx) = f(x_o) + dxf'(x_o) + \frac{dx^2}{2!}f''(x_o) + \frac{dx^3}{3!}f'''(x_o) + \ldots \tag{1.4}$$

If dx is small then higher order term of dx will be much smaller and hence (1.4) can be approximated as (1.5) or (1.6)

$$f(x_o + dx) = f(x_o) + dxf'(x_o) = y_o + dxf'(x_o) \tag{1.5}$$

$$dy = f(x_o + dx) - y_o = dxf'(x_o) \tag{1.6}$$

(1.6) defines a linear relation between small signal variation in x to small signal variation in y. Thus a nonlinear relationship can be simplified as linear relationship for analyzing small signal variations. The relationship shown in (1.6) also called as small signal equivalent of corresponding nonlinear equation. This concept is very useful in analyzing the behaviour of nonlinear elements for small signal change where nonlinear element is replaced by its equivalent small signal linear model.

Example: The Fig.1.6 shows a circuit which uses a nonlinear element diode whose I-V characteristics are as given in (1.2). The circuit is excited using one large signal DC source V_A and one small signal AC source v_a. When the small signal excitation is zero the current through diode is I_D and output voltage is V_D Draw its small signal equivalent circuit?

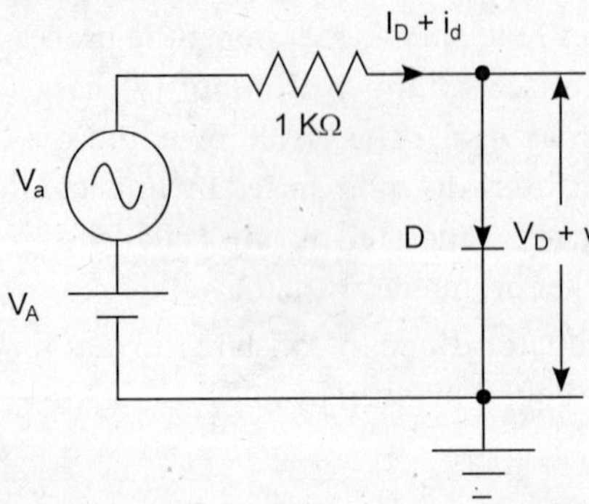

Fig.1.6: Circuit diagram having nonlinear element (Diode)

Solution: Using (1.6)

$$i_d = v_d \frac{dI}{dv}|_{I_D} \tag{1.7}$$

Again using (1.2)

$$\frac{dI}{dv} = \frac{I_s}{nV_T} e^{\frac{V}{nV_T}} \approx \frac{I}{nV_T} \tag{1.8}$$

Hence using (1.7) and (1.8)

$$i_d = v_d \frac{I_A}{nV_T} \tag{1.9}$$

The relationship given by (1.9) is a linear relationship which can be modelled using a resistor having equivalent resistance of r_d. This resistance is defined as dynamic resistance of the diode. It is defined in (1.10).

$$r_d = \frac{nV_T}{I_A} \tag{1.10}$$

Hence the equivalent small signal circuit is as shown in Fig.1.7.

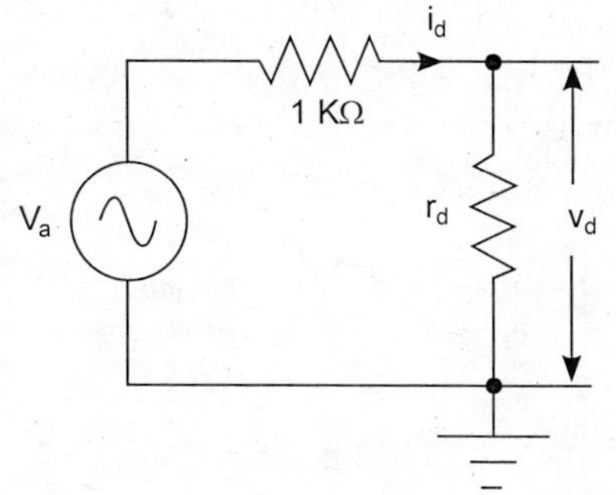

Fig.1.7: Small signal linear equivalent of circuit shown in Fig.1.6

1.5.3 Classification of Diodes based on its Application

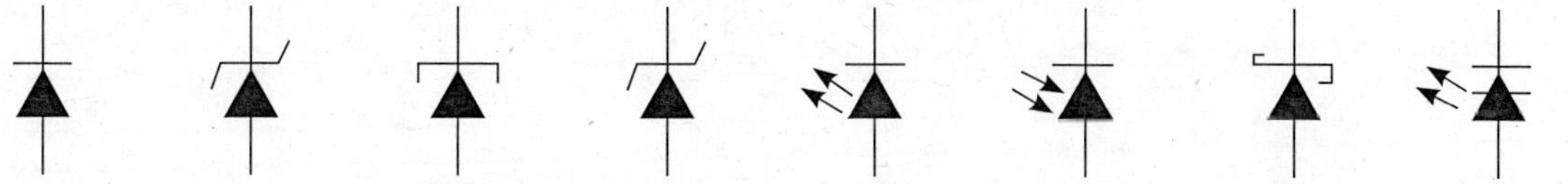

Fig.1.8: Symbols of different types of diodes

Based on its application, diodes are categorised in several different categories. These different types of diodes along with its symbols are listed in Fig.1.8. The direction of arrow in these symbol depicts the direction in which current can flow in forward bias. The arrowhead points toward the cathode signifying that current in diode flows only from anode to cathode and not vice versa. The details about the operation and working principle of these diodes are discussed below.

1. **P-N Junction diode:** A P-N junction diode is formed by bringing p-type semiconductor in contact with an n-type semiconductor. Its operation has been discussed in previous section. In a p-n junction diode, p-type semiconductor forms anode while n-type semiconductor forms cathode of the diode.
2. **Avalanche diode:** It is a special type of p-n junction diode which is designed to have avalanche breakdown at a specific reverse biased voltage. A regular p-n junction diode may get damaged due to avalanche breakdown. However, the junction of avalanche diode is designed in such a way that it avoids concentration of charge carriers preventing the hot spots. This is done by ensuring that avalanche breakdown is uniform across the junction which also results in fixed voltage for different currents.
3. **Tunnel diode:** Tunnel diode is a semiconductor diode designed using p-n junction where semiconductors are heavily doped. The heavy doping of p and n-type semiconductor results in alignment of valence band of p-type semiconductor with conduction band of n-type semiconductor. This allows tunnelling of electron from conduction band of n-type semiconductor to valence band of p-type semiconductor. This tunnelling of electron results in current during small forward bias voltage as shown in Fig.1.9. This tunnelling current increases as the bias voltage is increased.

However, beyond a certain forward bias voltage the two bands start getting misaligned and thus current decreases with any further increase in forward bias voltage. This phenomenon results in negative resistance. Tunnel diodes are useful in designing oscillator and amplifiers and also used to generate high frequency noise source.

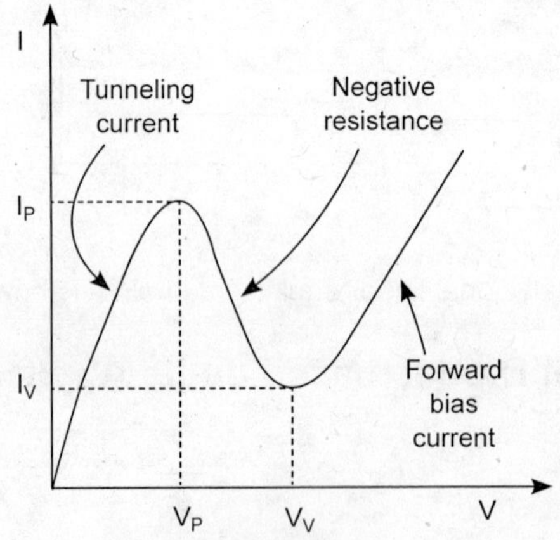

Fig.1.9: I-V characteristic of a tunnel diode

4. **Zener diode:** Zener diodes are heavily doped p-n junction diode. During reverse bias most of the applied voltage across Zener diode actually gets applied across depletion region which is very thin due to heavy doping. This results in a very high electric field in depletion region. This high electric field results in breaking of covalent bond and thus releases free charge. However these charges do not get enough time to accelerate to further boost the process as the depletion region is very thin. Hence, breakdown is mostly due to tunnelling and not due to impact ionization of electron-hole pair which is the case with avalanche breakdown. This breakdown is first noticed by Clarence Zener and hence termed as Zener breakdown. The temperature coefficient of Zener breakdown is negative which is different than avalanche breakdown which has positive temperature coefficient. The Zener diodes are used in voltage regulation. The Zener breakdown voltage is lower than avalanche breakdown voltage. The characteristic of a Zener diode is similar to the characteristic shown in Fig. 1.5 where during forward bias Zener diode act like a normal p-n junction diode. However during reverse bias when voltage across it reaches to Zener breakdown voltage then it enters into Zener breakdown and voltage across it becomes almost constant even if current changes.

5. **Light Emitting diode:** An LED is a semiconductor diode which emits light when it is forward biased. During forward bias, holes in p-type semiconductor move towards n-side which has plenty of electrons and electrons in n-type semiconductor move towards p-side which has plenty of holes and thus some of it get recombined in this process.

   The recombination results in release of energy which can either be thermal energy (Phonons) or electromagnetic energy in form of light (Photons) depending upon the types of semiconductor. The class of semiconductor where minimum of conduction band is directly above the maximum of valance band is known as direct semiconductor. The direct semiconductor results in release of photon during recombination and thus are used in fabrication of LED. The class of semiconductor where minimum of conduction band is not directly above the maximum of valance band is known as indirect semiconductor. Indirect semiconductor results in release of phonons. Si and Ge are example of an indirect semiconductor while GaAs and InAs are example of direct semiconductor.

6. **Photodiode:** Photodiode is a semiconductor diode which when exposed to photon having energy higher than its bandgap generates electron hole pairs and thus results in an electric current. It either operates in photovoltaic mode where it is biased at zero potential or in photoconductive mode where it is reverse biased. In some recent application it is also biased at avalanche breakdown voltage to have a greater sensitivity.
7. **Schottky diode:** A metal semiconductor junction results in a Schottky diode. Schottky diode offers a fast switching and has a low barrier potential. Typical metal used in Schottky diode are molybdenum, chromium, platinum and semiconductor is n-type silicon. The barrier potential of Silicon p-n junction diode is around 0.6 to 0.7V while barrier potential in Schottky diode is only around 0.15-0.45V.
8. **Laser diode:** A laser diode is a semiconductor diode which electrically has a PIN structure. In a PIN structure an intrinsic semiconductor layer is sandwiched between p-type semiconductor and n-type semiconductor. This intrinsic region is the active region of laser diode where charge carriers are confined to get recombined and thus generate photon in phase.

1.6 APPLICATIONS OF DIODE

Didoes are used in several real life applications. Some of these applications are discussed below.

1.6.1 Rectification

The behaviour of allowing the current flow in one direction is also called as rectification and hence diodes are also called as rectifier. Diodes are used in converting alternating current (AC) signal into direct current (DC) signal using this property. The circuits which perform rectification are called rectifier circuits. Following are some commonly used rectifier circuits.

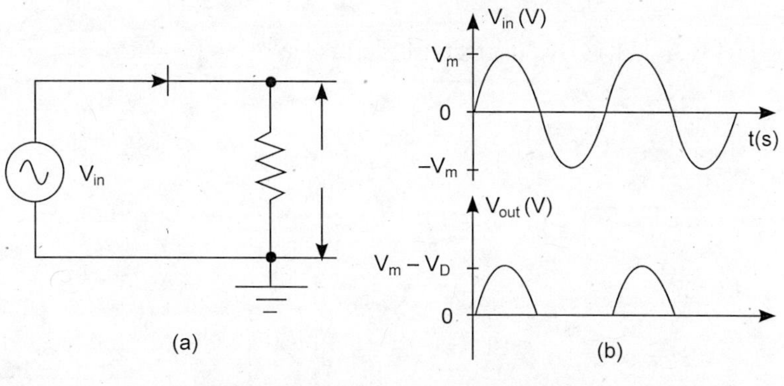

Fig.1.10: Half wave rectifier circuit

1. **Half wave rectifier:** Fig.1.10 (a) shows a half wave rectifier circuit which uses one p-n junction diode. For the positive half cycle of input signal (V_{in}) diode is in forward bias and hence it conduct while for the negative half cycle of input the diode is in reverse bias and hence it does not conduct. If the effective resistance of diode is much smaller than load resistance then input signal appears at output as it is during its positive half cycle however output remains at ground potential during its negative half cycle. Thus output (V_{out}) is a half wave rectified version of V_{in}. In this discussion effect of barrier potential is not considered. If we consider the effect of barrier potential then V_{out} will be V_{in}- V_D when Vin > V_D and it will be zero otherwise as shown in Fig.1.10 (b).

   The rectifier is characterized using several parameters. One such parameter is rectifier efficiency. Rectifier efficiency is defined as the ratio of output DC power to input AC power as expressed in (1.11). It is denoted using symbol η.

$$\text{Rectifier efficiency } (\eta) = \frac{P_{dc}}{P_{ac}} \tag{1.11}$$

   Rectifier efficiency of a half wave rectifier is calculated using (1.12),

$$\eta = \frac{P_{dc}}{P_{ac}} = \frac{\left(\frac{I_m}{\pi}\right)^2 R_L}{\left(\frac{I_m}{2}\right)^2 (R_L + r_f)} \tag{1.12}$$

   Where is the forward resistance of diode. Considering forward resistance of diode to be zero η becomes 40.6%.

   Another parameter which is used to characterize rectifier is ripple factor. Ripple factor is defined as the rms value of AC component V_{ac} to the rms value of DC component V_{dc} in rectifier output as given in (1.13). It is expressed using symbol Γ

$$\Gamma = \frac{V_{ac}}{V_{dc}} \tag{1.13}$$

   The rms voltage at rectifier output is defined as

$$V_{rms}^2 = V_{ac}^2 + V_{dc}^2 \tag{1.14}$$

   Hence using (1.13) and (1.14), ripple factor is defined as given by (1.15).

$$\Gamma = \sqrt{\left(\frac{V_{rms}}{V_{dc}}\right)^2 - 1} \tag{1.15}$$

   The ripple factor of a half wave rectifier is calculated as follows using (1.15),

$$\Gamma = \sqrt{\left(\frac{V_{rms}}{V_{dc}}\right)^2 - 1} = \sqrt{\left(\frac{\frac{V_m}{2}}{\frac{V_m}{\pi}}\right)^2 - 1} = 1.21 \tag{1.16}$$

Another parameter to characterize rectifier is peak inverse voltage (PIV). The PIV is defined as the maximum voltage experienced by diode when it is in reverse bias. It is important to know PIV of a diode in a circuit to ensure its safe operation. For a half wave rectifier the PIV is V_m.

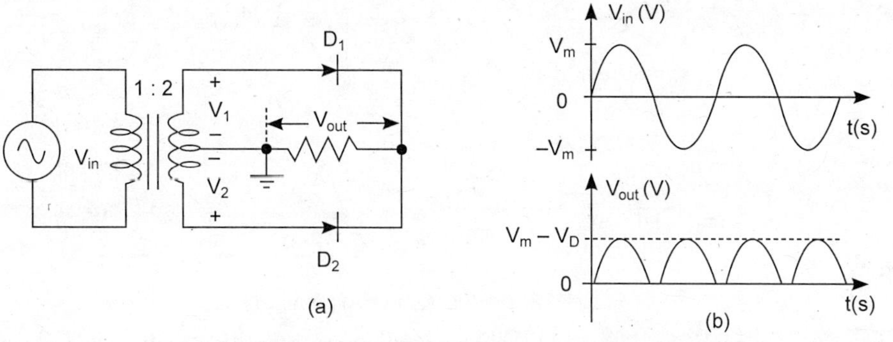

Fig. 1.11: Centre tapped full wave rectifier circuit

2. **Centre-tapped Full wave rectifier:** Full wave rectifier circuit can be designed in two possible ways. One of the configuration is shown in Fig.1.11(a) which uses two diode and a centre tapped transformer to design a full wave rectifier circuit. This circuit works as follows: The output of the centre tapped transformer generate two equal and out of phase version of input signal V_1 and V_2. Both of these signal are connected to output load in parallel using two diode D_1 and D_2 respectively. During positive half cycle of input signal V_{in}, the didoe D_1 become forward biased while D_2 is reverse biased and hence V_1 is connected to output. During negative half cycle of input signal V_{in}, the diode D_2 become forward biased while D_1 is reverse biased and hence V_2 is connected to output. Thus, output is connected to a version of input signal for both polarity of input signal. The output waveform of this circuit is shown in Fig. 1.11 (b) where output is $V_{in} - V_D$ when $V_{in} \geq |V_D|$ and zero otherwise.

   Rectifier efficiency of a full wave rectifier is given by (1.17)

$$\eta = \frac{P_{dc}}{P_{ac}} = \frac{\left(\frac{2I_m}{\pi}\right)^2 R_L}{\left(\frac{I_m}{\sqrt{2}}\right)^2 (R_L + r_f)} \tag{1.17}$$

   Where r_f is the forward resistance of diode. Considering forward resistance of diode to be zero η becomes 81.2%.

   Further, the ripple factor of a half wave rectifier is given using (1.18)

$$\Gamma = \sqrt{\left(\frac{V_{rms}}{V_{dc}}\right)^2 - 1} = \sqrt{\left(\frac{\frac{V_m}{\sqrt{2}}}{\frac{2V_m}{\pi}}\right)^2 - 1} = 0.48 \tag{1.18}$$

The PIV for a centre tapped full wave rectifier is $2V_m$.

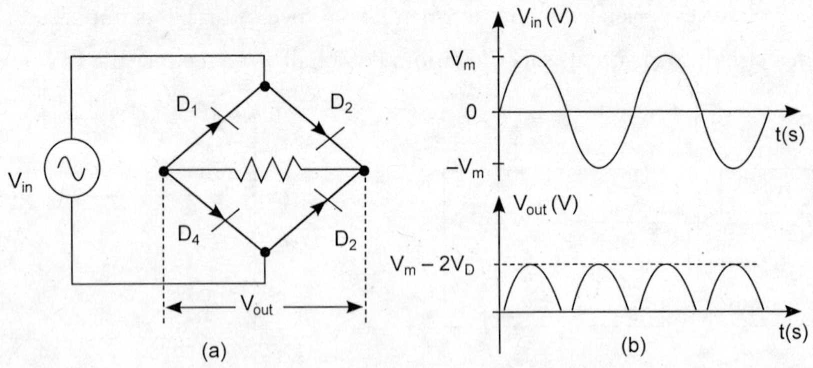

Fig. 1.12: Bridge full wave rectifier

3. **Full wave bridge rectifier:** Fig. 1.12 (a) shows another configuration to achieve full wave rectifier. This circuit uses a bridge configuration of 4 diodes and hence it is known as bridge full wave rectifier. It works as follows: During positive half cycle of input signal V_{in}, diode D_1 and D_3 becomes forward biased while D_2 and D_4 remain reverse biased and hence input is connected output load R_L with same polarity. During negative half cycle of input signal V_{in}, diode D_2 and D_4 becomes forward biased while D_1 and D_3 remain reverse biased and hence input is connected output load R_L with opposite polarity. Thus input signal always appears at output with positive polarity as shown in Fig.1.12(b). The output is $V_{in} - 2V_D$ when $V_{in} \geq |V_D|$ as two forward biased diode comes in signal path and zero otherwise.

   The ripple factor and rectification efficiency of full wave bridge rectifier is same as centre tapped full wave rectifier. However, PIV of full wave bridge rectifier is V_m only.

The output of rectifier circuit is unipolar but has AC components in it and hence it need to be filtered to get the DC output. A filter is a circuit which select a part of frequency response while rejecting other and hence it has a specific frequency transformation behaviour. A filter which select DC components while rejecting high frequency AC component comes under the category of low pass filter. Following are the possible filter circuit which can be used at the output of rectifier circuits.

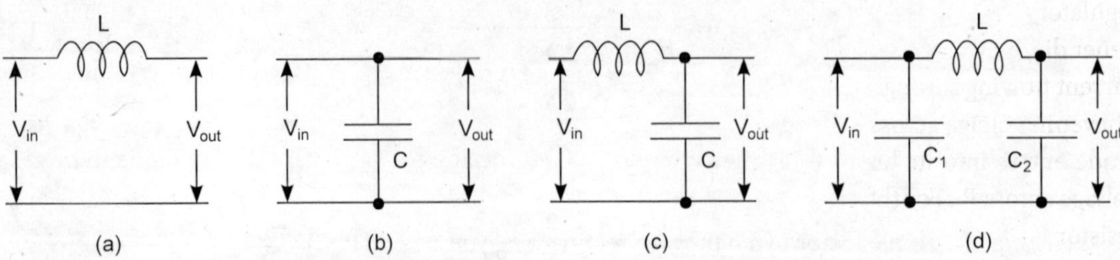

Fig. 1.13: (a) L Filter (b) C Filter (c) LC Filter (d) PI Filter

1. **L Filter:** L filter uses a series inductor between input and output port as shown in Fig.1.13(a). The inductor allows low frequency components while rejects high frequency components. Hence using a series inductor between output of rectifier and load reduces AC components in its output while maintaining the amplitude of DC components. It works as follows, when the output of rectifier

circuit passes through series inductor it offers high resistance to AC components thus blocking it while low resistance to DC components thus allowing it. Hence output of the filter circuit has primarily DC components. The L filters are effective in removing high frequency noise but it can be bulky as inductor occupies relatively larger area.

2. **C Filter:** C filter uses a shunt capacitor between input and output port. The capacitor allows high frequency components while rejects low frequency components. Hence using a shunt capacitor at output of rectifier in parallel to load as shown in Fig. 1.13 (b) reduces AC components in output while maintaining the amplitude of DC components. It is helpful in removing ripple at the output of the rectifier. The capacitors are compact and cheaper and hence C filters are relatively cheaper and more compact. However, it is not very effective in removing very high frequency components and additional filtering stages are required for it.
3. **LC filter:** Using series inductor along with shunt capacitor as shown in Fig. 1.13 (c) allows a higher order filtering. Further, the ripple factor of LC filter is almost independent of the load resistance while it is proportional to load resistance in L-filter and inversely proportional to load resistance in C-filter. This makes it attractive to use LC filter.
4. **PI filter:** PI filters are designed by arranging three passive components in a shape of Greek letter PI as shown in Fig.1.13(d). It can either be lowpass or high-pass depending upon the nature of circuit components used. The Fig.1.13(d) shows a low pass PI filter where a C filter is cascaded with an LC filter. It is also called capacitor input LC filter. The input capacitor is selected to offer low reactance to AC components to reject it. Hence most of the filtering is offered by this capacitor. The LC filter offers an additional filtering and hence filtering offered by CLC filter is much better than LC filter alone. The output voltage of PI filter can be very high with minimum voltage droop. It has very good ripple reduction feature however its regulation is poor.

1.6.2 Voltage Regulator Using Zener Diode

The circuit in Fig.1.14 is an example circuit where Zener diode is used to regulate the output voltage to Zener breakdown voltage V_z. When the voltage across Zener diode reaches to its Zener breakdown voltage then it enters into Zener breakdown where the voltage across it becomes almost constant even if current varies. Hence if Zener diode is biased around this operating point then it can act as a voltage regulator. The input to the circuit is a DC supply which is larger than V_z. If the Thevenin voltage across Zener diode is smaller than Zener breakdown voltage V_z then Zener diode is in reverse bias region where current flowing through it is small and can be considered zero. In this case I_Z is 0 and $I_S = I_L$. However, if Thevenin voltage across Zener diode is equal to or greater than Zener breakdown voltage V_z then Zener diode enters into its breakdown region and voltage across it becomes V_z by dropping the remaining voltage across R_S. For the case as shown in Fig.1.14, this condition becomes true when voltage across load resistor R_L becomes equal to Zener breakdown voltage V_Z. For this to happen the minimum load current $I_{L,\min}$ needed is defined using (1.19)

$$I_{L,\min} = \frac{V_{o,\min}}{R_L} = \frac{V_Z}{R_L} \tag{1.19}$$

Where $V_{o,\min}$ is the minimum voltage across load for Zener diode to enter into breakdown region which is equal to Zener breakdown voltage V_Z.

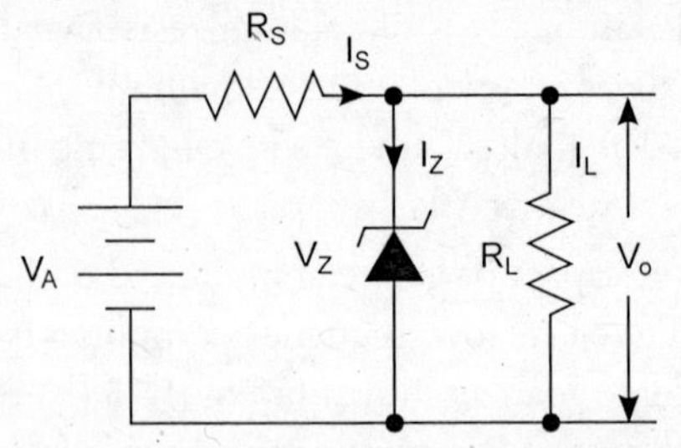

Fig. 1.14: Circuit diagram of voltage regulator using Zener diode

Example: Calculate the current IS, ID and IL shown in Fig.1.14 if R_s = 1kΩ, R_L = 1kΩ and V_A = 10, V_Z = 4. What is the value of output voltage?

Ans: The Thevenin voltage across diode is given using (1.20)

$$V_T = V_A \frac{R_L}{R_S + R_L} = 10 \times \frac{1}{2} = 5 \tag{1.20}$$

As diode is in reverse bias and Thevenin voltage across diode is higher than its Zener breakdown voltage. Hence diode will be operating in Zener breakdown region. Hence voltage across Zener diode which is also the output voltage will be 4V as shown in (1.21).

$$V_o = V_Z = 4\ V \tag{1.21}$$

so now I_L is given using (1.22),

$$I_S = \frac{V_A - V_Z}{1k\Omega} = \frac{6V}{1k\Omega} = 6mA \tag{1.22}$$

and I_L is given using (1.23),

$$I_L = \frac{V_Z}{1k\Omega} = \frac{4V}{1k\Omega} = 4mA \tag{1.23}$$

and hence Zener current I_Z is given using (1.24),

$$I_Z = I_s - I_L = 6\text{ mA} - 4\text{ mA} = 2\text{ mA} \tag{1.24}$$

1.6.3 Clipper Circuits

The rectification behaviour of diode is used to selectively shape a part of signal waveform and thus clipping a portion of it.

Series positive clipper circuit: Fig. 1.15(a) shows a series positive clipper circuit. It works as follows: During positive half cycle of input signal the diode is reverse biased and hence the output signal is grounded while during negative half cycle of input signal the diode becomes forward biased and hence output is same as input signal. In the above discussion effect of barrier potential is not considered. If we consider the effect of barrier potential then output V_o is given by (1.25).

$$V_0 = \begin{cases} 0 & (V_{in} > -V_D) \\ V_{in} - V_D & (V_{in} \le -V_D) \end{cases} \tag{1.25}$$

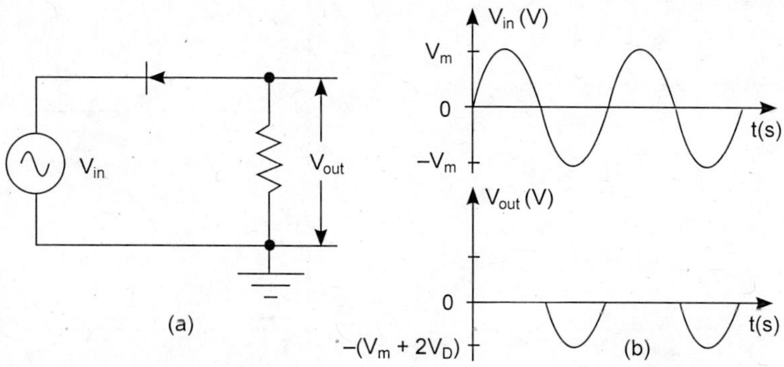

Fig. 1.15: (a) Circuit diagram of series positive clipper circuit (b) Input and output waveforms

Series negative clipper circuit: Fig. 1.16(a) shows a series positive clipper circuit. It works as follows: During positive half cycle of input signal the diode get forward biased and hence the output signal is same as input signal while during negative half cycle of input signal the diode becomes reverse biased and hence output is grounded. In the above discussion effect of barrier potential is not considered. If we consider the effect of barrier potential then output is given by (1.26)

$$V_0 = \begin{cases} V_{in} - V_D & (V_{in} \geq V_D) \\ 0 & (V_{in} < V_D) \end{cases} \quad (1.26)$$

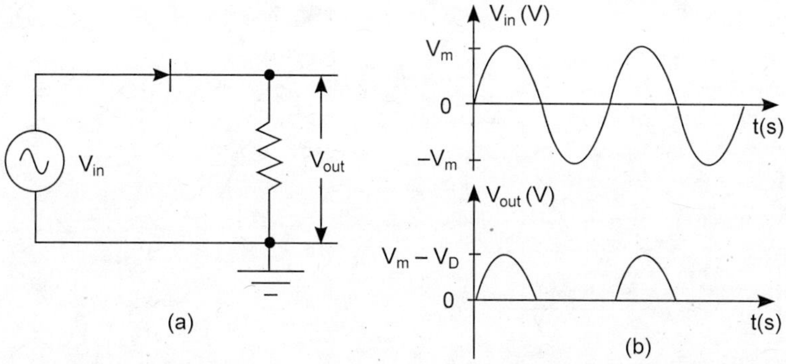

Fig. 1.16: (a) Circuit diagram of series negative clipper circuit (b) Input and output waveforms

Shunt positive clipper circuit: Fig. 1.17(a) shows a shunt negative clipper circuit. It works as follows: During positive half cycle of input signal the diode is reverse biased and hence the output signal is same as input signal while during negative half cycle of input signal the diode becomes forward biased and hence output is grounded. In the above discussion effect of barrier potential is not considered. If we consider the effect of barrier potential then output is given by (1.27)

$$V_o = \begin{cases} 0 & (V_{in} > V_D) \\ V_{in} & (V_{in} \leq V_D) \end{cases} \quad (1.27)$$

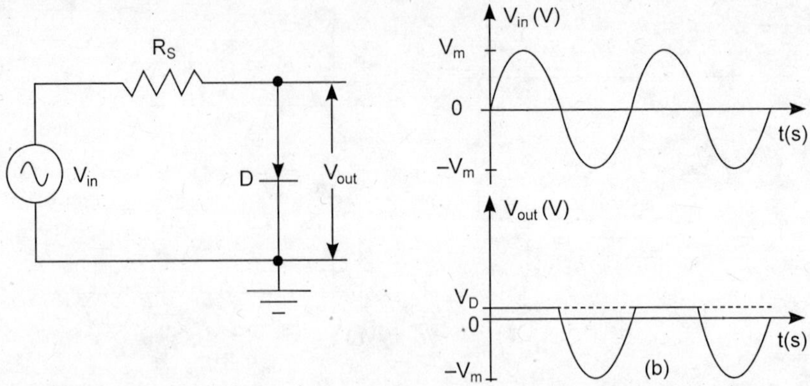

Fig. 1.17: (a) Circuit diagram of shunt positive clipper circuit (b) Input and output waveforms

Shunt negative clipper circuit: Fig. 1.18 (a) shows a shunt negative clipper circuit. It works as follows: During positive half cycle of input signal the diode is reverse biased and hence the output signal is same as input signal while during negative half cycle of input signal the diode becomes forward biased and hence output is grounded. In the above discussion effect of barrier potential is not considered. If we consider the effect of barrier potential then output is given by (1.28)

$$V_o = \begin{cases} V_{in} & (V_{in} \geq -V_D) \\ -V_D & (V_{in} < -V_D) \end{cases} \tag{1.28}$$

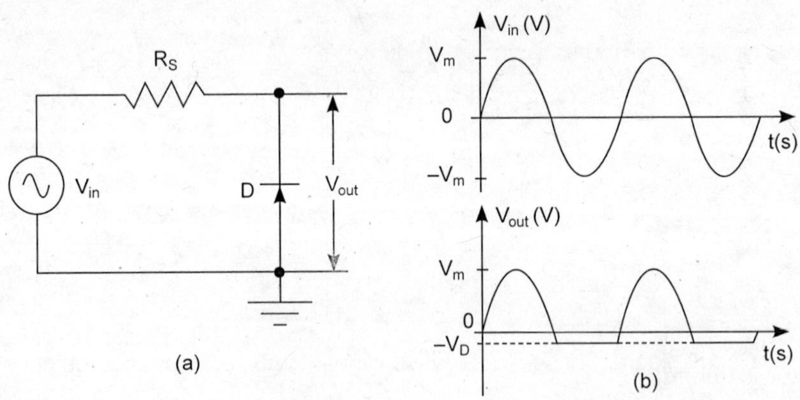

Fig. 1.18: (a) Circuit diagram of shunt negative clipper circuit (b) Input and output waveforms

If a diode is used in series with a signal source then clipping threshold changes and clipper circuit discussed above becomes biased clipper circuit. The following example discusses one such problem.

Example: Plot the output waveform for the circuit shown in Fig. 1.19 (a) where waveform of input signal is shown in Fig. 1.19 (b). Assume diode as an ideal diode with zero cut-in voltage.

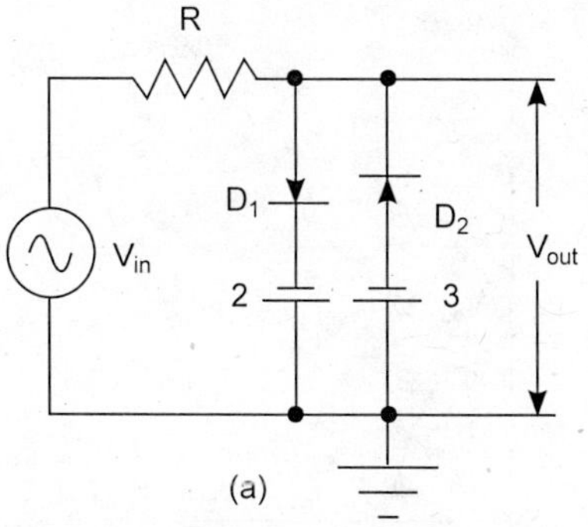

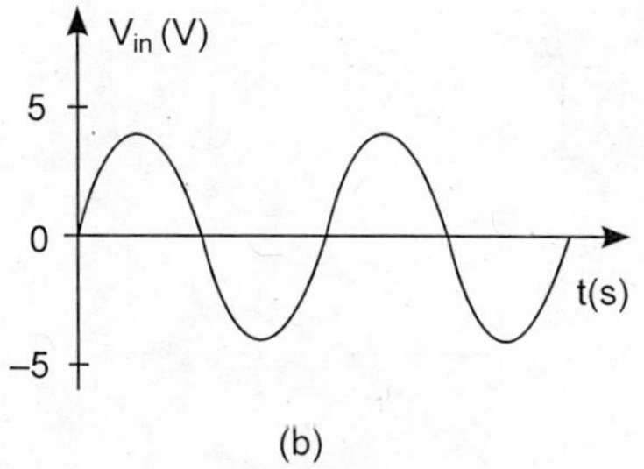

Fig.1.19: (a) Circuit diagram (b) Waveform of input signal V_{in}

Solution: For $V_{in} \geq 2$, D_1 is in forward biased and hence output will get clipped to 2V. Similarly for $V_{in} \leq -3$, D_2 becomes forward biased and hence output gets clipped to $-3V$. For $2 > V_{in} > -3$, both the diodes are reverse biased and hence input appears at output as it is. The output waveform of this circuit is given in Fig.1.20.

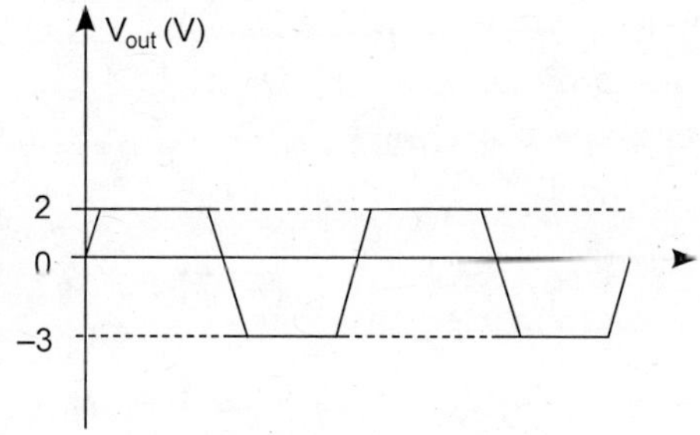

Fig. 1.20: Waveform of output voltage V_{out} for the circuit shown in Fig.1.19 (a)

1.6.4 Clamping Circuits

Clamping circuits are used to change the DC bias of a input signal. It uses a capacitor which is charged using a diode when the diode is forward biased and used as a battery source to shift the DC bias when the diode is reverse biased. Following are different types of clamping circuits:

Positive clamping circuit: Fig. 1.21 (a) shows a positive clamping circuit. It works as follows: Assume that the capacitor is initially discharged. As the negative half cycle of input signal begins, the diode D becomes forward biased and charges capacitor C to V_{in}. When the input signal reaches to its minimum potential the capacitor gets charged to $-V_m$. After this as the input start rising again the diode becomes reverse biased as voltage across it becomes negative. In this case the output voltage is a combination of input voltage and capacitor voltage which goes to a minimum of 0 and hence diode will never become forward biased again. In the above discussion effect of barrier potential is not considered. If we consider the effect of barrier potential then output is given using (1.29) and its waveform is shown in Fig. 1.21 (b).

$$V_o = V_{in} + V_c = V_{in} + V_m - V_D \tag{1.29}$$

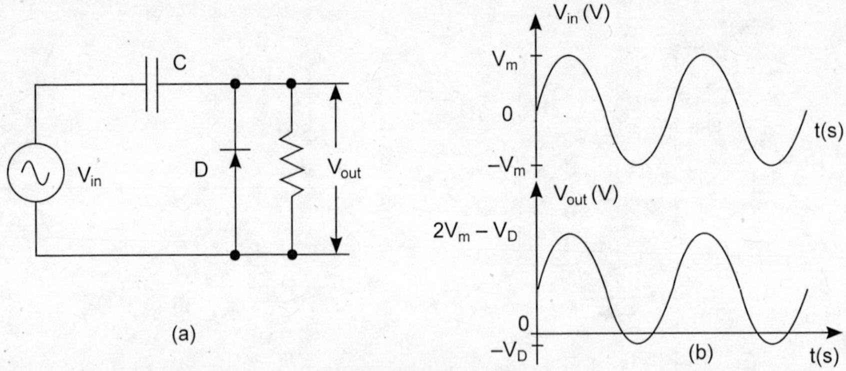

Fig.1.21: (a) Circuit diagram of positive clamper circuit (b) Input and output waveforms

Negative clamping circuit: Fig. 1.22 (a) shows a negative clamping circuit. It works as follows: Assume that the capacitor is initially discharged and hence capacitor start charging as the input rise. Now when the input reaches to its peak the capacitor is charged to V_M. As the negative half cycle of input signal begins, the diode D becomes forward biased and charges capacitor C to V_{in}. Now if input start decreasing, then output across diode which is a combination of input voltage and capacitor voltage becomes negative and hence diode becomes reverse biased. Here after voltage across diodes never becomes positive and hence output is always $V_{in} - V_m$. In the above discussion effect of barrier potential is not considered. If we consider the effect of barrier potential then output voltage is given by (1.30) as shown in Fig. 1.22 (b)

$$V_o = V_{in} + V_c = V_{in} - V_m + V_D \tag{1.30}$$

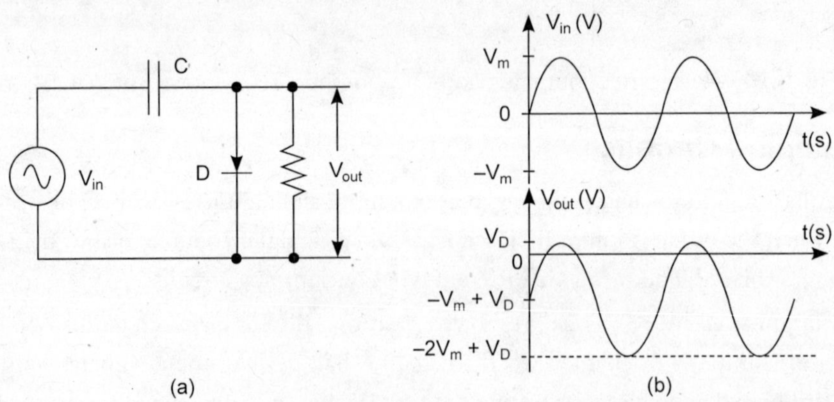

Fig.1.22: (a) Circuit diagram of negative clamper circuit (b) Input and output waveforms

Again using a bias source in series with diode changes the magnitude of voltage through which capacitor gets charges and thus shift in DC bias gets modified. This will be discussed in following example.

Example: The circuit in Fig.1.23(a) is excited using input waveform shown in Fig.1.23(b). Find and plot its output waveform if $V_M > V_A$.

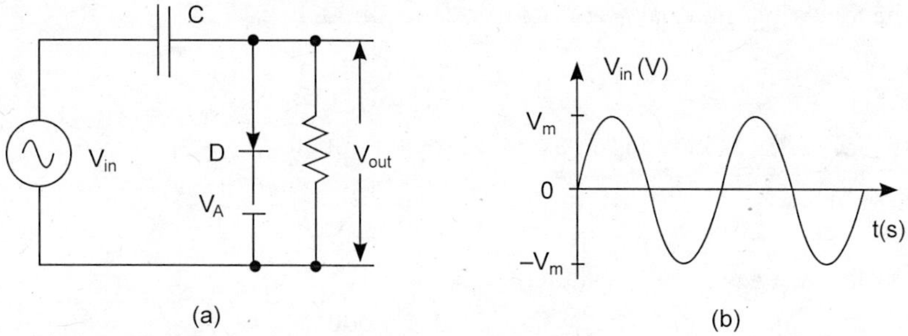

Fig.1.23: (a) Circuit diagram (b) Waveform of input signal V_{in}

Solution: Assume capacitor is initially discharged. If we start our analysis at a time when input signal is at its peak then the diode will become forward biased as V_M is higher than V_A. In this case capacitor get charged to($V_M - V_A$). and output at this time is V_A. As the input start decreasing the diode will become reverse biased and hence output becomes $V_{in} - (V_M - V_A)$. The diode thereafter remains reverse biased. The output waveform is shown in Fig.1.24.

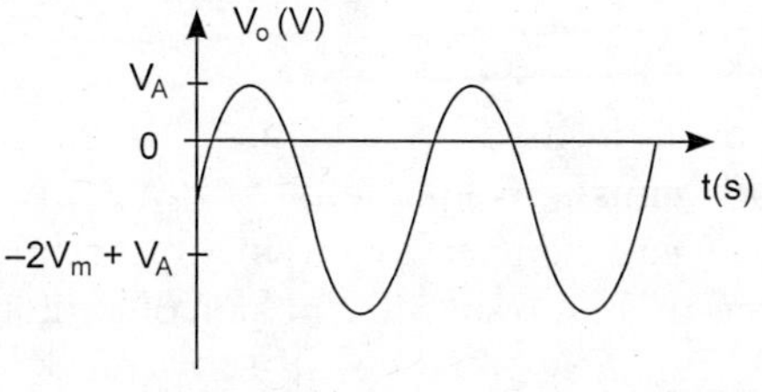

Fig.1.24: Waveform of output voltage V_o

UNIT SUMMARY

This unit provides an introduction of semiconductor and its properties. It also introduces the concept of intrinsic and extrinsic semiconductor. It further develops the concept of p-n junction and give details about its operation. The chapter provides a brief overview of different types of diodes and application of diodes. It also provides an introduction of different filter architectures which are used in rectifiers.

EXERCISES

Multiple Choice Questions

1. Which of the following material has least bandgap?
   (a) Metals
   (b) Semiconductors
   (c) Insulators
2. Which of the following material is a semiconductor material?
   (a) Sodium
   (b) Iron
   (c) Silicon
   (d) glass

3. Which of the following material is an intrinsic semiconductor?
   (a) Sodium (b) GaAs
   (c) n-type Si (d) p-type Si
4. Which of the following material is an extrinsic semiconductor?
   (a) Sodium (b) GaAs
   (c) Si (d) p-type Si
5. Which of the following are nonlinear element?
   (a) Resistor (b) Diode
   (c) Capacitor (d) Inductor
6. Which of the following is an active element?
   (a) Diode (b) Ideal voltage source
   (c) Resistor (d) Inductor
7. Which of the following is unilateral element?
   (a) Resistor (b) Voltage source
   (c) Diode (d) Capacitor
8. Which of the following statement is true?
   (a) LEDs are designed using indirect semiconductor.
   (b) Recombination results in emission of photons in direct semiconductor.
   (c) Recombination results in emission of phonons in direct semiconductor.
   (d) LED when biased in reverse bias results in emission of photons.
9. Which the following value closely depicts the rectification efficiency of a full wave rectifier?
   (a) 1.21 (b) 0.82
   (c) 0.46 (d) 0.48
10. Which of the following value closely depicts the ripple factor of a half wave rectifier?
   (a) 1.21 (b) 0.82
   (c) 0.46 (d) 0.48
11. Which of the following diode has least cut-in voltage?
   (a) Zener diode (b) LED
   (c) Schottky Diode (d) Photodiode
12. Which of the following diode are used for regulation?
   (a) Zener diode (b) Photodiode
   (c) Laser diode (d) LED

ANSWERS															
1.	(a)	2.	(c)	3.	(b)	4.	(d)	5.	(b)	6.	(b)	7.	(c)	8.	(b)
9.	(b)	10.	(a)	11.	(c)	12.	(a)								

Short Answer Type Questions

1. What is the difference between metal, semiconductor and insulator?
2. What is the difference between intrinsic and extrinsic semiconductor?

3. What are holes? What is the polarity of charge on holes?
4. What is diode? Is it an active or a passive device?
5. Explain how diode is a unidirectional device?
6. What is depletion region?
7. What is rectifier? Which device can be used as a rectifier?
8. Which type of diode emits light? Which semiconductors are used to fabricate this device?
9. What is the difference between light emitting diode and photodiode? Which bias should be used to operate these devices?
10. Name the diode which is used as a regulator?
11. What is direct semiconductor? Name the applications which requires direct semiconductors?
12. Which diode shows a negative resistance in its I-V characteristic?

Long Answer Type Questions

1. Explain the working of a p-n junction diode? What is depletion region and built in potential in p-n junction diode?
2. What is the definition of small signal? What are important points to consider while drawing small signal equivalent of a circuit?
3. Compare half wave rectifier, full wave rectifier and bridge rectifier circuits.
4. What is Zener breakdown? How it is different from avalanche breakdown?
5. What is the use of filter in rectifier circuit? Describe and compare different types of filter used in rectifier circuits?
6. Discuss all different types of clipper and clamper circuit? What are its applications?

Numerical Problems

1. An AC source is having rms value of 10V. Find the average value of DC output when (a) Half wave rectifier (b) Full wave rectifier (c) Bridge rectifier is used. Also find the PIV of each diode.
2. In above problem if forward resistance of diode is assumed to be around 5 ohm and if load is 500 ohm then find the peak current through diodes.

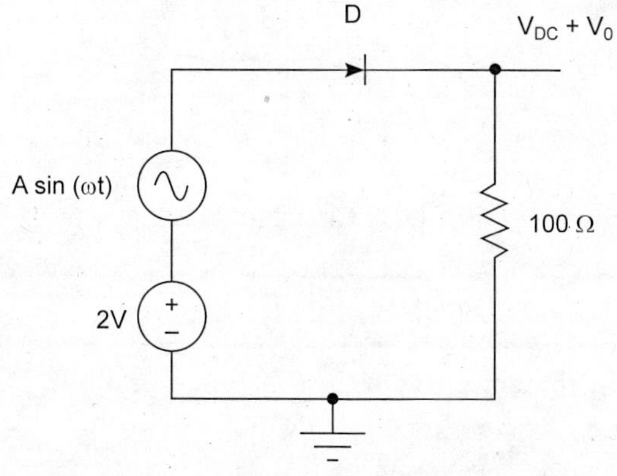

Fig. 1.25: Circuit for Numerical type Q.3

3. A 2V DC source is used to bias the diode circuit as shown in Fig.1.25. An AC source is connected in series with DC bias which acts as a small signal source. Here magnitude of AC source is A=20mV and frequency of AC source is 50Hz. The output of the circuit is a combination of DC output V_{DC} and small signal output v_o. One can consider diode as a battery of 0.7 V in forward bias. Find the small signal resistance of diode. Also find out V_{DC} and v_o.
4. For the circuit shown in Fig.1.17(a), input voltage source is a sinusoidal wave of amplitude 1V and frequency 50Hz. The diode is assumed to be of short circuit in its forward bias and open circuit in its reverse bias. Plot the waveform of its output voltage. Redo the exercise if diode has a drop of 0.7V in its forward bias.
5. For the circuit shown in Fig.1.23(a), input voltage source is a sinusoidal wave of amplitude 6V and frequency 50Hz and DC source V_A = 2V. The diode is assumed to be of short circuit in its forward bias and open circuit in its reverse bias. Plot the waveform of its output voltage. Redo the exercise if diode has a drop of 0.7V in its forward bias.
6. Design a clipper circuit using diodes to generate the signal shown in Fig.1.26(b) from the signal shown in Fig.1.26(a).

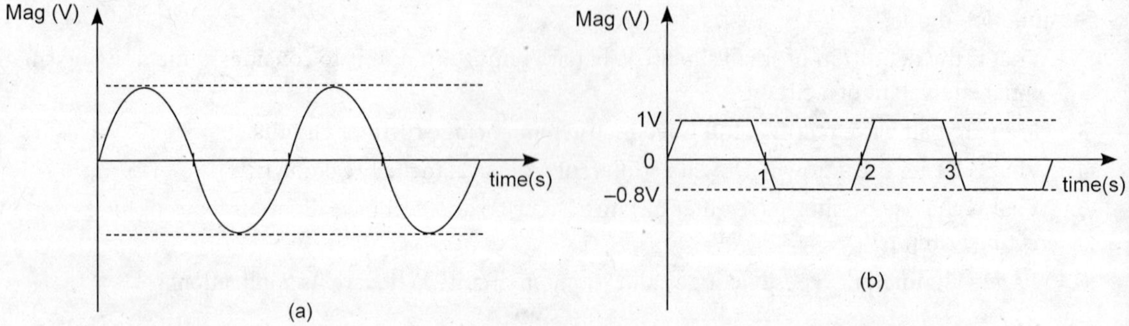

Fig.1.26: (a) Input Waveform (b) Output waveform, for clipper circuit in Q.N1.6

7. Find I, I_L and I_Z in the circuit shown in Fig.1.27.

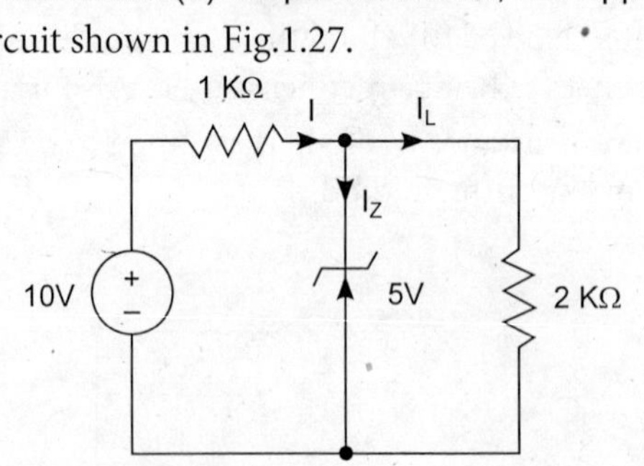

Fig.1.27: Circuit for Q.N1.7

PRACTICAL

Experiment-1.1: Plot the V-I characteristic of the P-N junction diode. Find its cut-in voltage.

Aim: Following are the aim of this experiment

1. Finding V-I characteristic of given diode.
2. Finding its cut-in voltage

Equipment and Components:

ક્રમાંક	Name of the component	Specification	Quantity
1	Diode	1N4007	1
2	Resistor	1kΩ	1
3	Resistor	10kΩ	1
4	Breadboard		1
5	Power supply	0-30V	1
6	Ammeter		1
7	Voltmeter		1
8.	Wires		

Procedure:

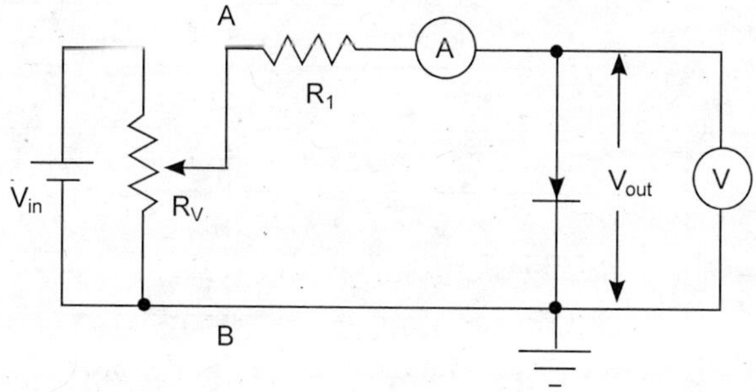

Fig.1.28: Circuit diagram for characterizing diode in forward bias

1. This experiment requires a variable power supply. One possible way to obtain a variable power supply is to use a DC source along with a variable resistor as shown in Fig.1.28. Check the power supply V_{in} and tune its voltage output between terminal A and B to around 0.3V. This value should be chosen slightly below the expected cut-in voltage of diode. If this value is not known then you can begin from 0V.
2. Note down the resolution and any error reading on Ammeter and voltmeter.
3. Arrange the components on breadboard and make all necessary connection to get the circuit shown in Fig.1.28.
4. Check the reading on ammeter, it should be close to zero.
5. Now start increasing the voltage output of power supply in steps of 50mV by tuning the variable resistor. Tabulate the readings of voltmeter and ammeter in Table.1
6. As the voltage across diode reaches to cut-in voltage the current through it start increasing quickly and hence reduce your step size to 10mV in this region. You will notice that current will not change much after some step as external resistor start limiting the current. However make sure you do not exceed the current and voltage rating of the device. You can stop if current does not show a change any further.

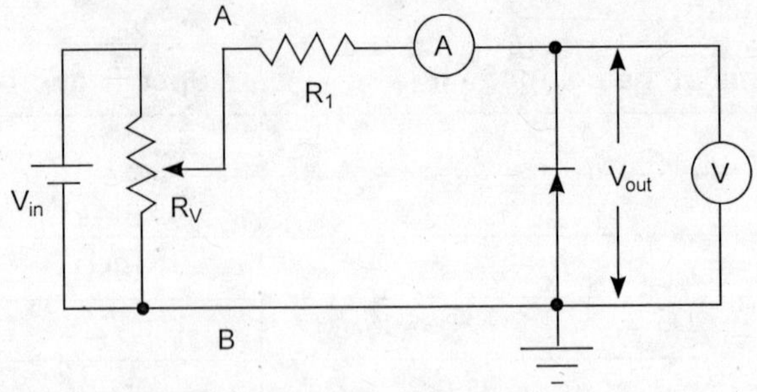

Fig.1.29: Circuit diagram for characterizing diode in reverse bias

7. Now change the polarity of diode as shown in Fig.1.29 to bias diode in reverse bias.
8. Set the voltage output of power supply at 1V and start increasing it in steps of 0.2V.
9. Plot the reading on same graph and estimate the reverse saturation current from it.

Observation

S.No.	Voltage	Current

Experiment-1.2: Plot the characteristic of a Zener diode. Find the breakdown voltage.

Aim:

1. To plot the characteristic of a Zener diode
2. To find the breakdown voltage.

Equipment and Components:

S.No.	Name of the component	Specification	Quantity
1	Zener Diode	1N4735	1
2	Resistor	1kΩ	1
3	Resistor	10kΩ	1
4	Breadboard		1
5	Power supply	0-30V	1
6	Ammeter		1
7	Voltmeter		1
8	Wires		

Procedure:

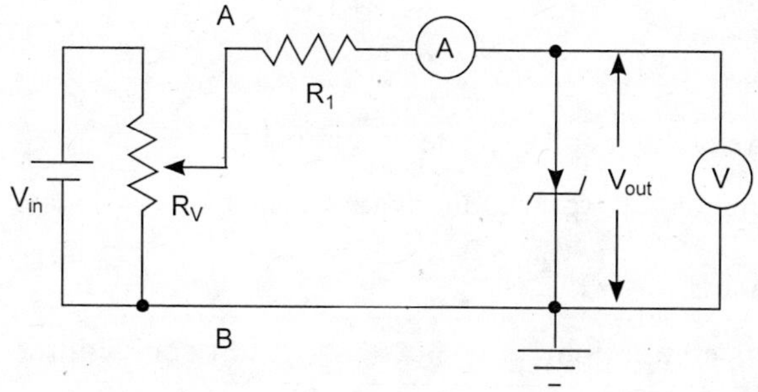

Fig.1.30: Circuit diagram for characterizing the Zener diode in forward bias

1. Check the power supply and tune its voltage output to around 0.3V. This value should be chosen slightly below the expected cut-in voltage of diode. If this value is not known then you can begin from 0V.
2. Note down the resolution and any error reading on Ammeter and voltmeter.
3. Arrange the components on breadboard and make all necessary connection to get the circuit shown in Fig.1.30.
4. Check the reading on ammeter, it should be close to zero.
5. Now start increasing the voltage output of power supply in steps of 50mV. Tabulate the readings of voltmeter and ammeter in Table.1
6. As the voltage across diode reaches to cut-in voltage the current through it start increasing quickly and hence reduce your step size to 10mV in this region. You will notice that current will not change much after some step as external resistor start limiting the current. However make sure you do not exceed the current and voltage rating of the device. You can stop if current does not show a change any further.

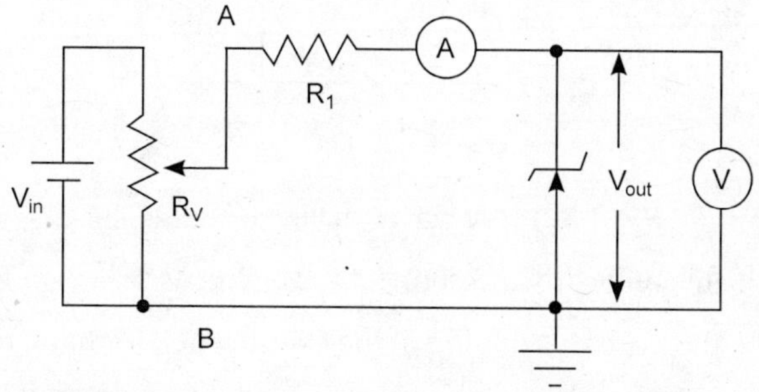

Fig.1.31: Circuit diagram for characterizing the Zener diode in reverse bias

7. Now make the connection as shown in Fig.1.31.
8. Set the voltage output of power supply at 1V and start increasing it in steps of 0.2V. As the diode will reach to Zener breakdown the current will increase sharply. Now increase voltage in steps of 0.1V.
9. Plot the reading on same graph and estimate the reverse saturation current from it.

Experiment-1.3: Regulation characteristic of half wave rectifier without filters and with filter. Compare its characteristic.

Aim:

1. To design a half wave rectifier and evaluate its output
2. To evaluate the output of half wave rectifier with C-Filter

Equipment and components:

S.No.	Name of the component	Specification	Quantity
1	Diode	1N4007	1
2	Signal source		1
3	Resistor	1kΩ,5kΩ 10kΩ	1 Each
4	Electrolytic capacitor		1 Each
5	Multimeter		1
6	Oscilloscope		1
7	Wires		

Procedure:

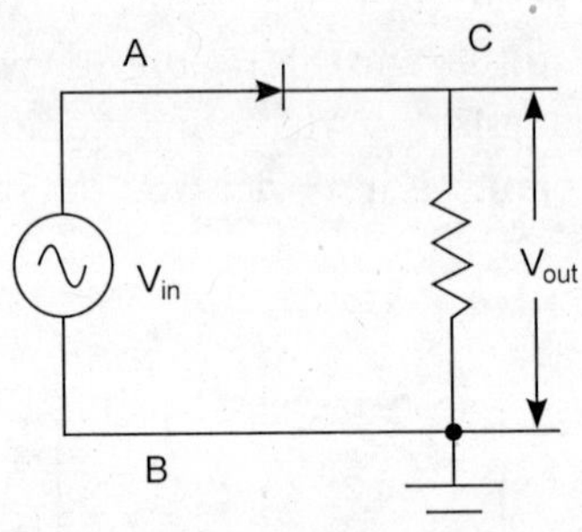

Fig.1.32: Half wave rectifier without filter

1. Arrange the components on board and connect the circuit as shown in Fig.1.32.
2. Apply a AC signal to the primary winding of the transformer. Measure the input AC voltage using multimeter on primary side. Also measure the AC voltage on secondary side of the transformer. Multiply the rms value with $\sqrt{2}$ to get the amplitude V_{max} of AC signal. The amplitude of DC output will be $\frac{max}{}$. Now measure the DC output voltage using multimeter. Compare it with calculated value.
3. Repeat the step 2 by changing the signal strength of input signal.

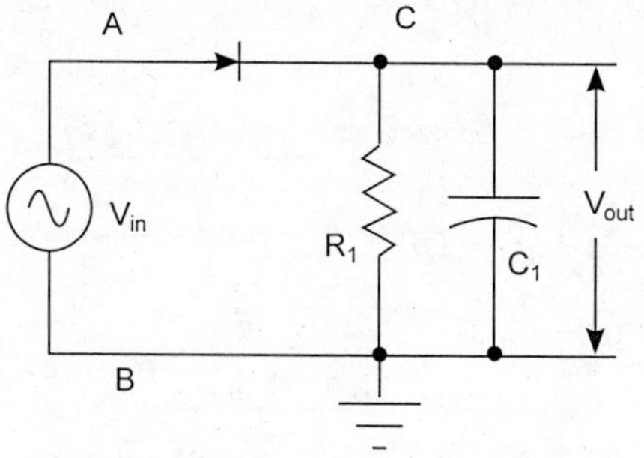

Fig.1.33: Half wave rectifier with filter

4. Now add the electrolytic capacitor C_1 as shown in Fig.1.33 such that positive terminal of the capacitor is connected to rectifier output and negative terminal of the capacitor is connected to ground. Connect the output of rectifier to oscilloscope to see the output waveform. Measure the ripple in output. Calculate ripple factor using expression $\Gamma = \sqrt{\left(\frac{V_{rms}}{V_{DC}}\right)^2 - 1}$.
5. Repeat the step 4 with different value of capacitor.

Experiment-1.4: Regulation characteristic of centre-tapped full wave rectifier without filters and with filter. Compare its characteristic.

Aim:

1. To design a full wave rectifier and evaluate its output
2. To evaluate the output of full wave rectifier with C-Filter

Equipment and components:

S.No.	Name of the component	Specification	Quantity
1	Diode	1N4007	2
2	Centre tapped transformer	9-0-9	1
2	Signal source		1
3	Resistor	1kΩ,5kΩ 10kΩ	1 Each
4	Electrolytic capacitor	22	1 Each
5	Multimeter		1
6	Oscilloscope		1
7	Wires		

Procedure:

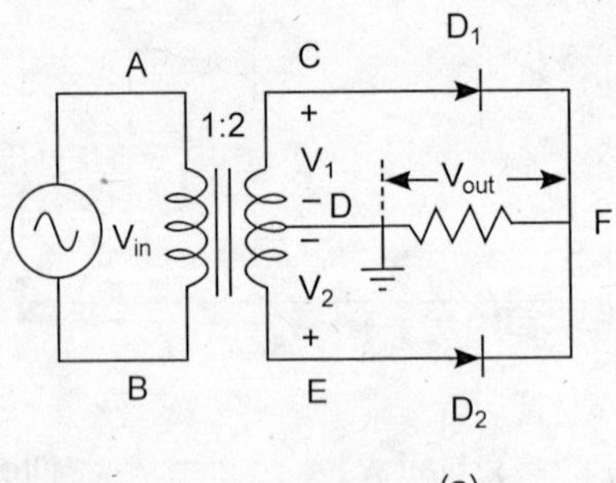

(a)

Fig.1.34: Full wave rectifier without filter

1. Arrange the components on board and connect the circuit as shown in Fig.1.34.
2. Apply a AC signal to the primary winding of the transformer. Measure the input AC voltage using multimeter on primary side. Also measure the AC voltage on secondary side of the transformer. Multiply the rms value with $\sqrt{2}$ to get the amplitude V_{max} of AC signal. The amplitude of DC output will be $\frac{V_{max}}{\pi}$. Now measure the DC output voltage using multimeter. Compare it with calculated value.
3. Repeat the step 2 by changing the signal strength of input signal.

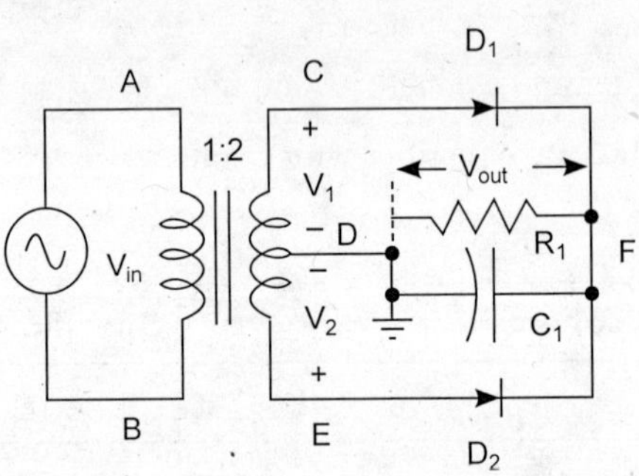

Fig.1.35: Full wave rectifier with filter

4. Now add the electrolytic capacitor as shown in Fig.1.35 such that positive terminal of the capacitor is connected to rectifier output and negative terminal of the capacitor is connected to ground. Connect the output of rectifier to oscilloscope to see the output waveform. Measure the ripple in output. Calculate ripple factor using expression $\Gamma = \sqrt{\left(\frac{V_{rms}}{V_{DC}}\right)^2 - 1}$.
5. Repeat the step 4 with different value of capacitor.

Experiment-1.5: Regulation characteristic of full wave bridge rectifier without filters and with filter. Compare its characteristic.

Aim:

1. To design a full wave rectifier and evaluate its output
2. To evaluate the output of full wave rectifier with C-Filter

Equipment and components:

S.No.	Name of the component	Specification	Quantity
1	Diode	1N4007	4
2	Signal source		1
3	Resistor	1kΩ, 5kΩ 10kΩ	1 Each
4	Electrolytic capacitor	22	1 Each
5	Multimeter		1
6	Oscilloscope		1
7	Wires		

Procedure:

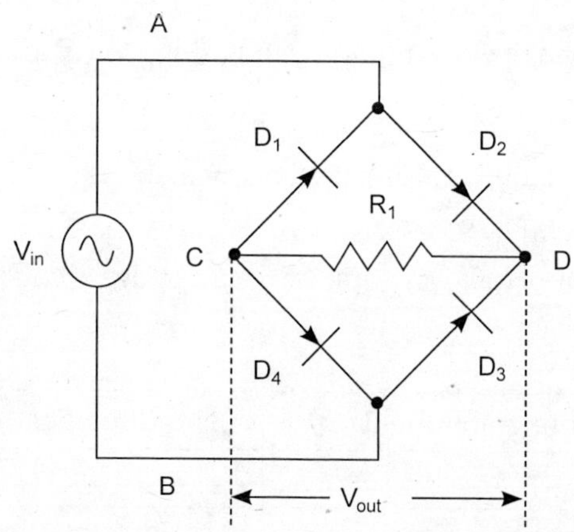

Fig.1.36: Full wave bridge rectifier without filter

1. Arrange the components on board and connect the circuit as shown in Fig.1.36.
2. Apply a AC signal to the circuit and measure the AC voltage on by connecting the multimeter across terminal A and B. Multiply the rms value with $\sqrt{2}$ to get the amplitude V_m of AC signal. The amplitude of DC output will be $\frac{V_m}{\pi}$. Now measure the DC output voltage using multimeter. Compare it with calculated value.
3. Repeat the step 2 by changing the signal strength of input signal.

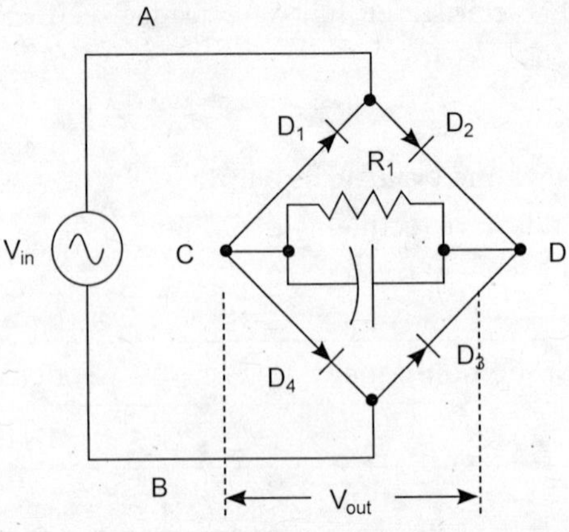

Fig.1.37: Full wave bridge rectifier with filter

5. Now add the electrolytic capacitor C_1 as shown in Fig.1.37 such that positive terminal of the capacitor is connected to rectifier output and negative terminal of the capacitor is connected to ground. Connect the output of rectifier to oscilloscope to see the output waveform. Measure the ripple in output. Calculate ripple factor using expression $\Gamma = \sqrt{\left(\frac{V_{rms}}{V_{DC}}\right)^2 - 1}$.

6. Repeat the step 4 with different value of capacitor.

Experiment-1.6: Simulation of half wave, full wave and bridge rectifier using simulation tool.

Aim: Following are the aim of this simulation

1. To gain an insight of simulation tool and design schematic of given circuit
2. To design schematic of circuits.
3. To simulate the behaviour of half wave, full wave and bridge rectifier rectifier

Equipment and Components:

S.No.	Name of the component	Specification	Quantity
1	Computer		
2	Spice simulator	LTspice or any other spice simulator	
3	PDK models	Any technology node such as 180nm	1

Theory related to simulation :

Simulation is a method to analyze the behaviour or performance of a circuit using a computer without realizing it on hardware. The circuit is described to a computer using languages such as SPICE which is an acronym for Simulation Program with an Integrated Circuit Emphasis. A sample SPICE program looks like following,

Title of the circuit

Description of the circuit (This part may extend up to several lines)

.MODEL STATEMENTS

ANALYSIS COMMANDS

OUTPUT COMMANDS

.END

Here first line of the program defines the title of the circuit. This line is not considered during compilation and printed directly to the display output. After this next few lines describes the circuit. This part of program is termed as netlist. Thereafter, the path of model files are provided using .MODEL keyword. The model files provide information about behaviour of device. This modelling can either be done in hardware description language for behavioural modelling or in SPICE for experimental models. These model files are included in netlist with .MODEL keyword to define the operation of device. The analysis commands are used to define different simulation methods to analyze the performance of circuit. For example operating point analysis is performed to obtain the information about the operating points of the circuit. The DC analysis is performed to analyze the variation in operating points if one of the circuit parameter varies. The transient analysis is performed to obtain the time domain behaviour of the circuit where a time domain input is applied to the circuit and its behaviour is obtained as a function of time. The AC analysis is a small signal analysis where the circuit is first linearized around the operating point and then this linear version of circuit is simulated across different frequency to obtain the frequency response of the circuit. The AC analysis is always linear so magnitude of applied input signal is not much important if one want to predict the gain of the system using this analysis. The noise analysis is also a small signal analysis where small signal version of circuit is analyzed with noise models of devices. There are many other simulation methods which are not much important for the experiment discussed here but its information can be easily obtained through the resources available online. Thereafter set of output commands are used to display results obtained from simulations.

Procedure:

1. Download a circuit simulator such as LT-Spice which can be downloaded at no cost from internet.
2. Launch the circuit simulator and create a new schematic file. Open this newly created schematic file.
3. Now the objective is to create a schematic of the circuit shown in Fig.1.32, Fig.1.34 and Fig.1.36. Let's first begin with circuit of Fig.1.32 Now, instantiate all the components used in this circuit. The components are instantiated by using Add component option available in LT spice.
4. Now connect the components to create the circuit shown in Fig.1.32.
5. The behaviour of the components are described using model file. The model file is a part of process development kit (PDK) which can be downloaded from website of respective foundry or supplier of the component. The model file is added by inserting a spice directive .INCLUDE <path> in the schematic window
6. Now again use SPICE directives to add the analysis commands to schematic window.

   The command to add operating point analysis is following.

   .op

   The command to add transient analysis is following

.TRAN STEP STOP START

Here STEP is the time step used by simulator to perform analysis and should be smaller than minimum expected event time. The STOP is the time up to which simulation is performed. The START is the time at which simulation begins.

7. Now simulate the circuit by clicking over the run simulation button.
8. Simulation results are available in *.raw file which can be opened using waveform window. The simulated data can then be seen by selecting appropriate data and plotting it. The small signal information is available in log files and not directly in waveform window.
9. Plot the results and save the waveforms.
10. Repeat the steps for circuit shown in Fig.1.34 and Fig.1.36.

KNOW MORE

Diode, is an early invention in the field of semiconductor which is still evolving. Different new structure of semiconductor diodes are continuously being developed to serve different applications. Detecting single photons using diode is an attractive application of it. It has become possible using single photon avalanche diode (SPAD). This Diode is biased well above its reverse biased breakdown voltage so that even a single photon is able to bring the diode into its avalanche state and thus provides a much higher sensitivity. A quenching circuit is used to take diode out of this state by lowering its bias voltage to a smaller value than its reverse biased breakdown voltage. The bias voltage is then again increased to make it able to detect another activity.

Dynamic QR Code for Further Reading

REFERENCES AND SUGGESTED READINGS

1. W. Shockley, "Electrons and Holes in Semiconductors" Princeton N.J.: Van Nostrand. 1950 pp.111-114.
2. Horowitz, Paul; Winfield Hill (1989). *The Art of Electronics, 2nd Ed*. London: Cambridge University Press. p. 44.
3. Acosta, Orlando N. "Zener diode-a protecting device against voltage transients." *IEEE Transactions on Industry and General Applications* 4 (1969): 481-488.
4. Shockley, William. "The Theory of p-n Junctions in Semiconductors and p-n Junction Transistors." *Bell system technical journal* 28, no. 3 (1949): 435-489.

2 Bipolar Junction Transistor (BJT)

UNIT SPECIFICS

Through this unit we have discussed the following aspects:

- Introduction to BJT operation
- DC Small signal modelling of BJT
- Analysis of different BJT configurations
- Designing different amplifier circuits using BJT
- High frequency analysis of BJT

The practical applications of the topics are discussed for generating further curiosity and creativity as well as improving problem solving capacity.

Besides giving a large number of multiple choice questions as well as questions of short and long answer types marked in two categories following lower and higher order of Bloom's taxonomy, assignments through a number of numerical problems, a list of references and suggested readings are given in the unit so that one can go through them for practice. It is important to note that for getting more information on various topics of interest some QR codes have been provided in different sections which can be scanned for relevant supportive knowledge.

After the related practical, based on the content, there is a "Know More" section. This section has been carefully designed so that the supplementary information provided in this part becomes beneficial for the users of the book. This section mainly highlights the initial activity, examples of some interesting facts, analogy, history of the development of the subject focusing the salient observations and finding, timelines starting from the development of the concerned topics up to the recent time, applications of the subject matter for our day-to-day real life or/and industrial applications on variety of aspects, case study related to environmental, sustainability, social and ethical issues whichever applicable, and finally inquisitiveness and curiosity topics of the unit.

RATIONALE

This chapter provides an introduction to the basic operation and working of Bipolar junction transistor(BJT). It discusses solid state physics behind it, provides a small signal model to analyse operation of BJT circuits when operated in small signal fashion. Different BJT configuration are introduced to provide a better understanding of circuits designed using BJT. Amplifiers designed using these different configurations are analyzed to compare its performance.

The chapter also provides high frequency model of BJT to analyze the behaviour of circuits designed using BJT when operated at high frequencies.

PRE-REQUISITES

Mathematics: Co-ordinate Systems (Class XII)

Physics: Mechanics (Class XII)

UNIT OUTCOMES

List of outcomes of this unit is as follows:

U2-O1: Understanding operation of BJT

U2-O2: Small signal modelling of BJT

U2-O3: To understand different BJT configuration

U2-O4: To learn design procedure of amplifier using BJT

U2-O5: High frequency models of BJT

Unit-2 Outcome	EXPECTED MAPPING WITH COURSE OUTCOMES (1 - Weak Correlation; 2 - Medium Correlation; 3 - Strong Correlation)				
	CO-1	CO-2	CO-3	CO-4	CO-5
U2-O1	3	3	2	1	2
U2-O2	3	3	3	2	2
U2-O3	2	3	3	2	3
U2-O4	1	3	3	2	3
U2-O5	2	3	3	1	2

2.1 INTRODUCTION

The diode discussed in previous chapter is a two terminal device. It allows current flow through it only when it is forward biased and thus it performs wave shaping. In this chapter we will discuss a three terminal device known as transistor which is useful in wave shaping, switching, amplification and many other applications. A transistor is an electronic device which controls the current flowing through it based on its control input. It can be thought of as a water tap where knob of the water tap is similar to its control input and flow of water is like current flowing through it. As the flow of water is controlled by the position of knob in water tap, in similar fashion control input in transistor controls the magnitude of current flowing through the transistor. It is categorised based on its control mechanism. The two major categories of transistors are Bipolar junction transistor (BJT) and Field effect transistors (FET).

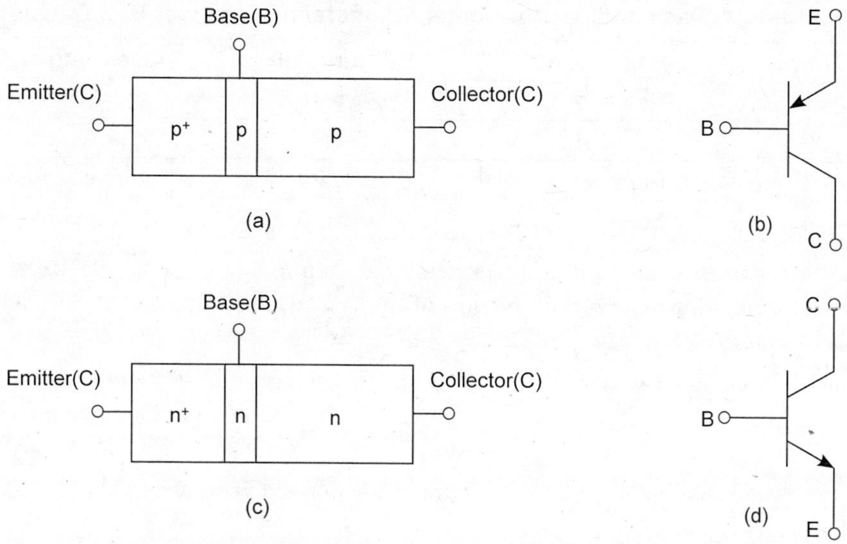

Fig. 2.1: (a) Structure of a p-n-p transistor (b) Symbol of a p n p transistor (c) Structure of a n-p-n transistor (d) Symbol of a n-p-n transistor

The BJT was invented by John Bardeen, Walter Brittain and William Shockley at Bell laboratories in 1948. It is named as a bipolar device because its operation involves both electrons (negatively charged) and holes (positively charged). It has two variants, p-n-p transistor and n-p-n transistor. The Fig.2.1(a) shows a structure of a p-n-p transistor and its symbol is shown in Fig.2.1(b). The dual of it will be a n-p-n transistor whose structure and symbol are shown in Fig.2.1(c) and Fig.2.1(d) respectively. These devices primarily have three regions, the leftmost region named as emitter is responsible for emitting the charge carriers. The emitter region is heavily doped so that it can provide sufficient number of charge carriers. The middle region is named as base. The base is kept thin and lightly doped so that charge carriers coming from emitter region quickly pass through this region without recombining with the opposite charge carriers present in the base region. Thus most of the charge carriers emitted from emitter are finally collected by the rightmost region which is known as collector region.

A transistor can also be thought of as a combination of two p-n junction diodes connected back to back and thereby forming a p-n-p transistor. However for transistor operation the middle sandwiched layer has to be very thin otherwise transistor action is not observed. This is why a transistor operation is not noticed when one attempts to connect two discrete p-n junction diode in back to back fashion. A transistor action refers to the amplifying action of the transistor. This will be discussed in detail in later part of this chapter.

2.2 OPERATING REGIONS OF BJT

BJTs have two p-n junctions, Emitter-Base junction (E-B) and Base-Collector junction (B-C). Thus there are four possible biasing scenarios which are listed in Table.2.1.

Table 2.1 Bias conditions for different operating region of BJT Transistors

S.No.	E-B junction	B-C junction	Operating region
1	Reverse	Reverse	Cut-off
2	Forward	Reverse	Active
3	Reverse	Forward	Reverse Active
4	Forward	Forward	Saturation

If both E-B and B-C junctions are reverse biased then to a large extent device is off as depletion region across the junction will prevent the flow of charge carriers. Hence this region of BJT operation is defined as cutoff region.

The second scenario is when E-B junction is forward biased and B-C junction is reverse biased. This region of operation is defined as active region. Here we discuss the operation of a p-n-p transistor. The dual of it will be applicable for a n-p-n transistor.

To make the discussion easy, terminal currents are defined as shown in the Fig.2.2. These currents are related by (2.1)

$$I_E = I_B + I_C \tag{2.1}$$

And hence collector current is given using

$$I_C = I_E - I_B \tag{2.2}$$

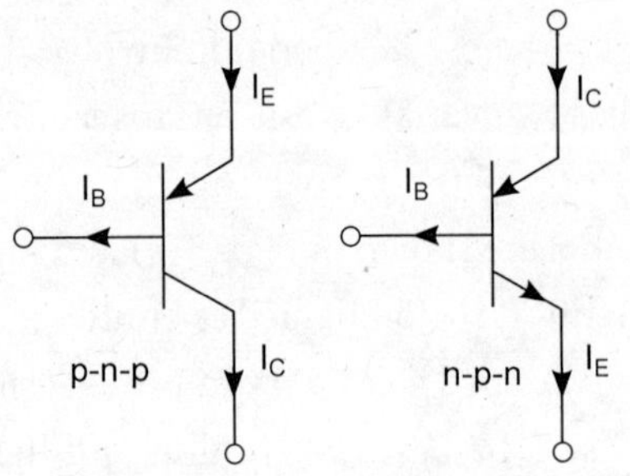

Fig. 2.2: Current terminology in p-n-p and n-p-n BJT transistor

Now if E-B junction is forward biased then I_E will be largely a diffusion current. This diffusion current has two components, the current due to movement of holes from emitter to base region and the current due to movement of electrons from base to emitter region as shown in Fig.2.3. However as emitter is heavily doped and hence diffusion current will be primarily due to movement of holes. Now when these holes reach to base which is of n-type then it start recombining with the electrons present in it. If the base is constructed very thin and lightly doped and further, B-C junction is reverse biased then most of these holes will quickly diffuse through the base region and will be swept away and collected in collector region as the depletion electric field in B-C junction favours its flow. Hence a significantly large amount of current will flow from emitter to collector region of the transistor. The net current which reaches to collector region is defined as collector current I_C. This current which flows through B-C junction is primarily contributed by movement of holes from n-type base region to p-type collector region. It is a function of V_{BE} and does not depend on V_{BC} to a large extent. The current in base terminal I_B is due to diffusion of electrons towards E-B junction, a part of it also recombines with holes coming from emitter. This current is generally much smaller compare to collector current. The common base current gain α_F is close to unity as β_F is a large number. This region of BJT operation is defined as active region as BJT can be used as an amplifier in this region of operation.

The ratio of collector current I_C to base current I_B in a BJT is an important DC parameter of BJT defined as common emitter current gain (β_F). Its expression is given in (2.3)

$$\beta_F = \frac{I_C}{I_B} \tag{2.3}$$

The ratio of collector current I_C to emitter current I_E of a BJT is defined as common base current gain (α_F). Its expression is given in (2.4)

$$\alpha_F = \frac{I_C}{I_E} = \frac{I_C}{I_C + I_B} = \frac{I_C / I_B}{I_C / I_B + 1} = \frac{\beta_F}{\beta_F + 1} \tag{2.4}$$

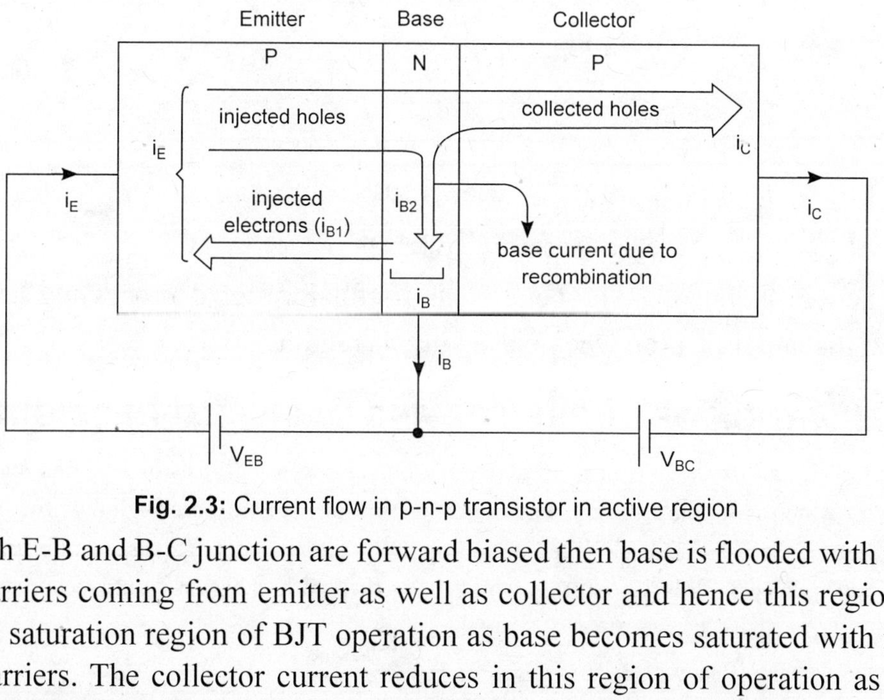

Fig. 2.3: Current flow in p-n-p transistor in active region

If both E-B and B-C junction are forward biased then base is flooded with minority charge carriers coming from emitter as well as collector and hence this region is also known as saturation region of BJT operation as base becomes saturated with minority charge carriers. The collector current reduces in this region of operation as forward biased B-C junction opposes the flow of holes coming from emitter region.

As BJT has a palindrome structure so if E-B junction is reverse biased and B-C junction is forward biased then a current will flow from collector to emitter which will be a function of V_{CE}. However, this current will be relatively smaller as collector is lightly doped and hence current gain will be much smaller in this case. This region of operation is defined as inverse active region.

In order to get a better understanding about the region of BJT operation Fig.2.3 shows plot of collector current vs collector base voltage V_{CB} in p-n-p transistor. The region in which V_{CB} is negative is the region where B-C junction is reverse biased. Further, if I_B is non-zero suggesting B-E junction is forward biased then this region is active region. As seen in the plot collector current is constant in active region suggesting it is independent of V_{CB} in active region. Now as V_{CB} becomes positive suggesting B-C junction becomes forward biased then collector current reduces as forward biased B-C junction does not favour the flow of holes emitted by emitter through B-C junction and thus collector current reduces. This region is saturation region.

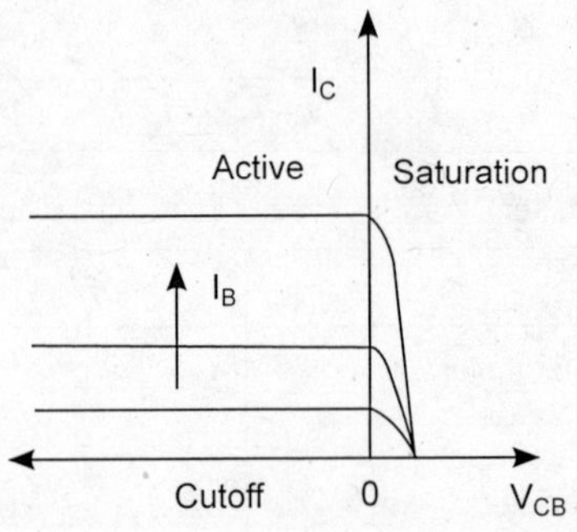

Fig. 2.4: characteristic of p-n-p transistor

The active region of BJT operation is useful for signal processing as BJT can provide signal amplification when biased in this region.

2.3 DERIVATION OF CURRENT'S AND GAIN'S EXPRESSIONS

In order to derive an expression of I_C in active region for a p-n-p transistor, one can start from the E-B junction which is forward biased. The concentration of holes in base region at the boundary of E-B depletion region $h(0)$ and B-C depletion region $h(W_B)$ are given using quasi equilibrium boundary condition or Shockley boundary condition where W_B is the width of base region. Its expression is given in (2.5) and (2.6)

$$h(0) = \frac{N_{iB}^2}{N_B}(e^{\frac{V_{BE}}{V_T}} - 1) \tag{2.5}$$

$$h(W_B) = \frac{N_{iB}^2}{N_B}\left(e^{\frac{V_{BC}}{V_T}} - 1\right) \approx -\frac{N_{iB}^2}{N_B} \approx 0 \tag{2.6}$$

Here N_{iB} is the intrinsic carrier concentration of base material and N_B is the base doping concentration. Now these holes which are minority carrier in base region results in diffusion current which finally flows to collector region as reverse biased B-C junction favours its flow. The charge carriers in the base are generated due to the diffusion of carriers across the forward bias E-B junction. Then these charge carriers after reaching the reverse bias electric field of B-C junctions are swept across to collector due to drift. Hence the number of charge carriers are still governed by diffusion across E-B junction only. This current which is actually collector current is given using the expression shown in (2.7) where A_E is cross sectional area of emitter, D_B is a constant.

$$I_c = \left| A_E q D_B \frac{dn}{dx} \right| = A_E q D_B \frac{h(0)}{W_B} = A_E q \frac{D_B}{W_B} \frac{N_{iB}^2}{N_B} \left(e^{\frac{V_{BE}}{V_T}} - 1 \right) \tag{2.7}$$

It is evident from this expression that collector current is a function of V_{BE} and it does not depend on V_{BC}.

Similarly electrons from base region move to emitter and results in a base current which is not desired but is a consequence of forward biased E-B junction. The expression of this current can also be derived in the similar way as collector current and is given using (2.8)

$$I_B = A_E q \frac{D_E}{W_E} \frac{N_{iE}^2}{N_E} \left(e^{\frac{V_{BE}}{V_T}} - 1 \right) \tag{2.8}$$

Where W_E is emitter length and D_E is a constant. The ratio of collector current to base current is defined using β_F which is given by (2.9)

$$\beta_F = \frac{I_c}{I_B} = \frac{D_B N_{iB}^2}{D_E N_{iE}^2} \frac{W_E N_E}{W_B N_B} \tag{2.9}$$

β_F is generally a large number because base current is much smaller than collector current. It is a most important DC parameter of the transistor and is also termed as common emitter current gain of the transistor. Its desirable value is around 100 or higher. As D,W and N_i do not change a lot so the only parameter which can be adjusted is N which is doping concentration. A large β_F is achieved by choosing a higher N_E and smaller N_B. Although choosing a smaller N_B may result in higher base resistance and can become a problem while operating transistor at higher current level or at higher frequencies.

2.4 EARLY EFFECT

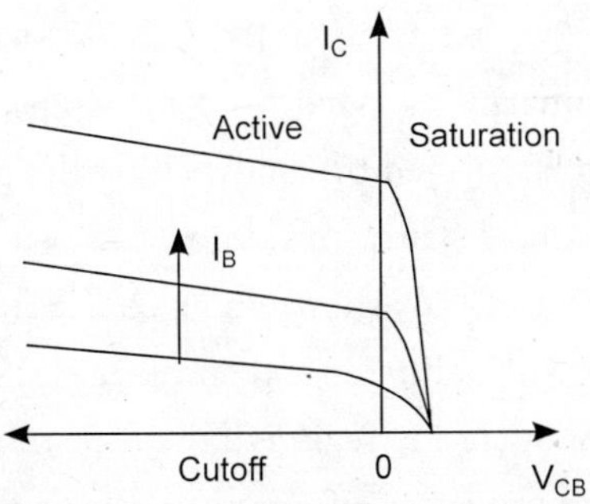

Fig. 2.5: Characteristic of a p-n-p transistor with early effect

Earlier we said that I_c in BJT is independent of V_{CB}. However, in reality I_c shows a finite non zero slope with respect to V_{CB} as shown in Fig.2.5. The reason behind this finite non-zero slope is base width modulation. The width of depletion region across B-C junction increases as V_{CB} becomes more negative and hence effective width of base region reduces which increases I_C. The characteristic of I_C when plotted versus V_{CE} also shows a similar behaviour as shown in Fig.2.6. In this characteristic if $I_C - V_{CE}$ curve is extrapolated then it intersect the x axis at the same point which is defined as early voltage V_A. Hence slope of transistor characteristic in active region is given by (2.10)

$$g_o = \frac{\partial I_C}{\partial V_{CE}} = \frac{I_C}{V_A} \tag{2.10}$$

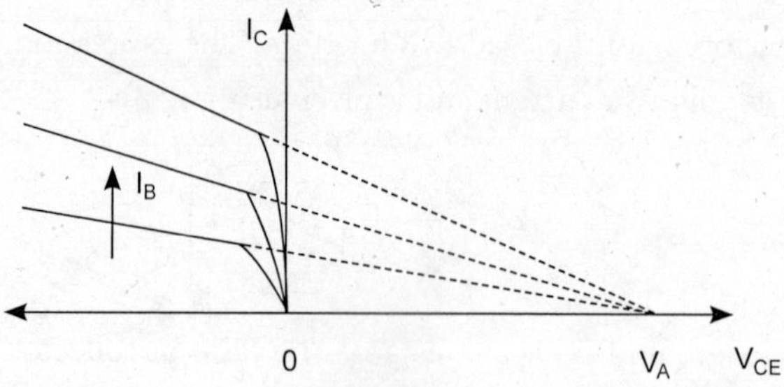

Fig. 2.6: Characteristic of p-n-p transistor with early effect showing early voltage

Dr. James Early who discovered early effect wrote in an anecdote that he discovered it while he was getting bored in a meeting and started thinking about how collector current can be a function of collector voltage so he wrote the relationship $I_C = \beta_F I_B$ and differentiated it with respect to collector voltage. The relationship then became $\frac{\partial I_C}{\partial V_C} = I_B \frac{\partial \beta_F}{\partial V_C}$ assuming base current I_B is constant. Now the question is how can β_F changes with collector voltage? As collector voltage is raised, B-C depletion layer thickens which effectively reduces base width and thus current gain increases. Quite obvious. Isn't it?

Further, relationship between base current and emitter current is found using (2.1) and (2.3). This is given by

$$I_E = I_B + I_C = I_B + \beta_F \cdot I_B = (1 + \beta_F) \cdot I_B \tag{2.11}$$

So far our discussion has primarily focussed on p-n-p transistor to explain the BJT operation. However a BJT transistor can also be formed by sandwiching the p-type semiconductor between two n-type semiconductor region. This new structure is known as n-p-n transistor and is dual of a p-n-p transistor. Its schematic and symbol is already shown in Fig.2.1(c) and Fig.2.1(d). Its operation is similar to the operation of p-n-p transistor with a difference that role of electron and hole is exchanged. The bias condition for operating in different operating region of the n-p-n transistor remains same as described in Table.2.1. The transconductance and speed of n-p-n transistor is higher than its p-n-p counterpart because of higher mobility of electrons than holes. Hence n-p-n transistors are more common in circuits designed using BJT transistors.

2.5 BIASING OF BJT

Now if one want to design a circuit using BJT transistor then the first step is to ensure the proper region of BJT operation. This becomes possible by maintaining specific currents and voltages through all devices. This process of establishing DC current through device (which in this case is BJT transistor) using additional circuit elements like voltage source, resistor etc is known as biasing. Fig.2.7 shows some possible ways of biasing a n-p-n transistor. The circuit shown in Fig.2.7 (a) uses a base resistor R_B to fix the base current of n-p-n transistor. Assuming a constant value of V_{BE} which is largely true because of exponential I-V characteristic of BJT, the value of I_B is given by (2.12)

$$I_B = \frac{V_{DD} - V_{BE}}{R_B} \tag{2.12}$$

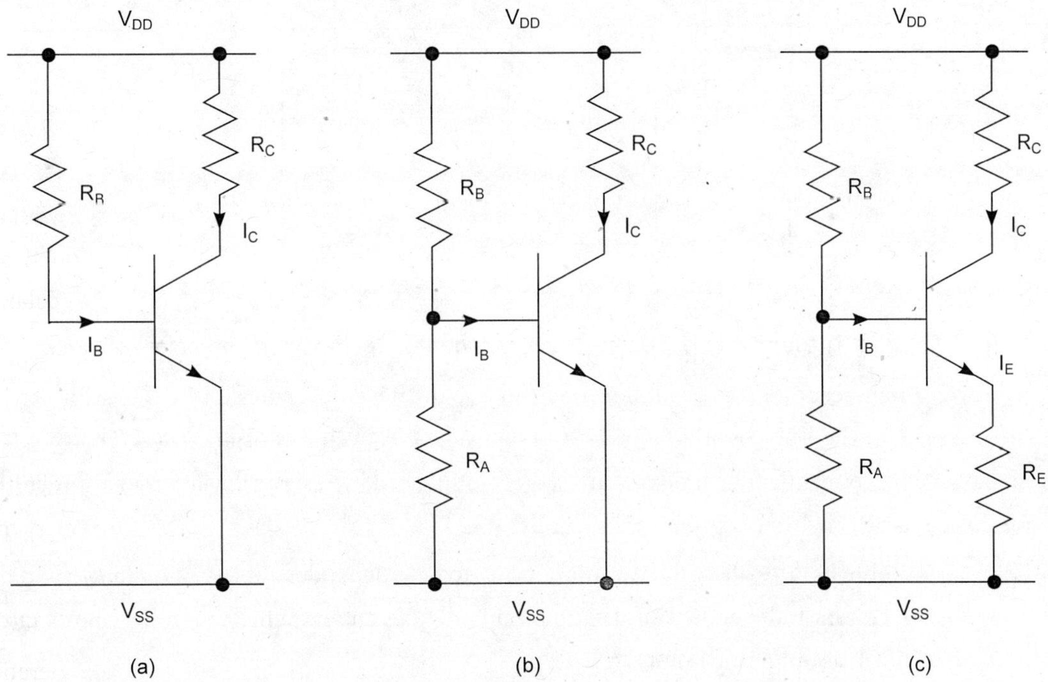

Fig. 2.7: Biasing of a n-p-n transistor (a) using base resistor (b) using resistive divider (c) emitter degeneration

However, bias current (I_C) is still a function of β_F as shown in (2.13) which is a process dependent parameter and hence this biasing is not robust across process.

$$I_C = \beta_F I_B = \beta_F \frac{V_{DD} - V_{BE}}{R_B} \tag{2.13}$$

Consider the biasing circuit shown in Fig.2.7 (b). It uses a resistor divider to fix the base emitter voltage V_{BE} while assuming base current is small. The base emitter voltage V_{BE} becomes $\frac{V_{DD} \cdot R_A}{(R_A + R_B)}$ if we assume base current much smaller to current flowing through the resistive divider. This can lead to collector current I_C being independent of β_F as shown in (2.14)

$$I_C = I_S(e^{\frac{V_{BE}}{V_T}} - 1) \approx I_S(e^{\frac{V_{BE}}{V_T}}) = I_S(e^{\frac{V_{DD}}{V_T}\frac{R_A}{(R_A+R_B)}}) \tag{2.14}$$

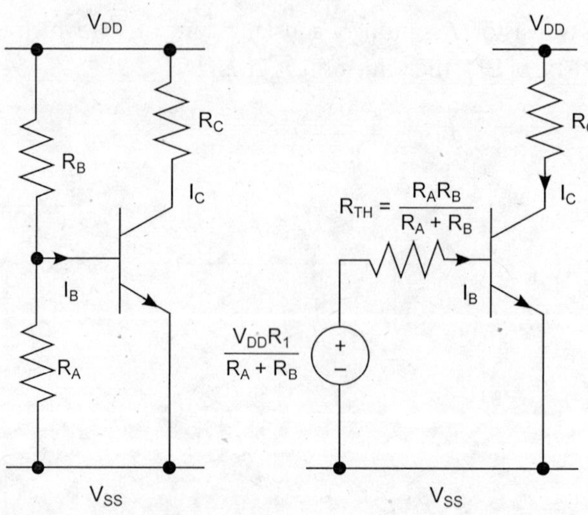

Fig. 2.8: Using Thevenin equivalent of resistor divider for finding operating points of circuit

However, in an accurate representation one can use a Thevenin equivalent of a resistor divider as shown in Fig.2.8 where V_{BE} is still a function of equivalent Thevenin resistance R_{TH}. Thus collector current having an exponential dependence on V_{BE} gets influenced by resistor deviation in a similar way. It make this circuit strongly susceptible to variations.

Using a resistor in emitter allow the variation in V_{BE} to be independent of deviation in R_{TH} and thus making collector current independent of it. The biasing defines the currents and voltages in a circuit which are also known as operating points.

BJT is a nonlinear device which is evident from its I-V characteristic shown in (2.14). Therefore, amplifier designed using it shows signal distortion when amplitude of input signal is large. However, if BJT is operated in its small signal range by restricting its input amplitude then its I-V characteristic becomes approximately linear which pave the way for designing distortion less linear amplifiers. The analysis of these circuits gets simplified on using linear equivalent model of BJT which is applicable in small signal range. These models are termed as small signal models of BJT

2.6 SMALL SIGNAL MODEL OF BJT

In order to develop the small signal model of BJT, Hybrid parameters or h-parameters which is a type of two port network parameters are used to model the behaviour of BJT. The two port parameters are preferred as it conveniently model amplifiers which are the most probable use case of BJT in a circuit. Among two port parameters, h-parameters closely model the behaviour of transistor and hence chosen to model it. In case of h-parameter input current and output voltage are chosen as independent variable and output current and input voltage are chosen as dependent variable. The Fig.2.9 shows the representation of a two port network using h-parameters. Here port (1-1') is input port and port (2-2') is an output port. By definition port is a pair of terminal where current at one terminal entering into network should be equal to current at other terminal exiting from the network.

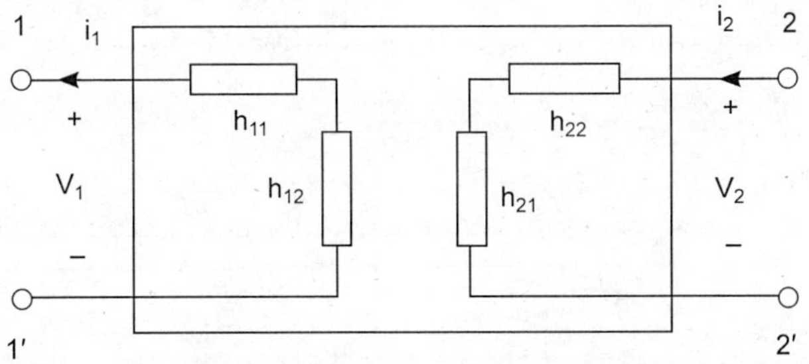

Fig. 2.9: Two-port h-parameter

The relationship between input and output voltages and current for 2-port h-parameter network is given using (2.15) and (2.16)

$$V_1 = h_{11}\, I_1 + h_{12}\, V_2 \tag{2.15}$$

$$I_2 = h_{21}\, I_1 + h_{22}\, V_2 \tag{2.16}$$

Where h parameters are defined as given in (2.17)

$$h_{11} = \frac{V_1}{I_1}\Big|_{V_2=0}, h_{12} = \frac{V_1}{V_2}\Big|_{I_1=0}, h_{21} = \frac{I_2}{I_1}\Big|_{V_2=0}, h_{22} = \frac{I_2}{V_2}\Big|_{I_1=0} \tag{2.17}$$

Here, h_{11} is input impedance of circuit, h_{12} is reverse voltage gain, h_{21} is forward current gain and h_{22} is output admittance. As the dimension of these parameters are not alike and hence it is termed as hybrid parameter or h-parameter. In the context of transistor h_{11} is termed as h_{ix}, h_{12} is termed as h_{rx}, $h21$ is termed as h_{fx} and h_{22} is termed as h_{ox} where x is substituted with BJT configuration such as e in case of common emitter configuration. These configurations will be discussed in later part of this chapter.

Generally in BJT h_{12} is relatively small and hence can be neglected. Then this h-parameter model is simplified with the hybrid-π model shown in Fig.2.10. The network parameters of these two models are related using expressions given in (2.18)

$$h_{11} = r_\pi, h_{12} = 0, h_{21} = \beta_F, h_{22} = \frac{1}{r_o} \tag{2.18}$$

The various parameter used in hybrid-π model are obtained by following the steps similar to (1.6).

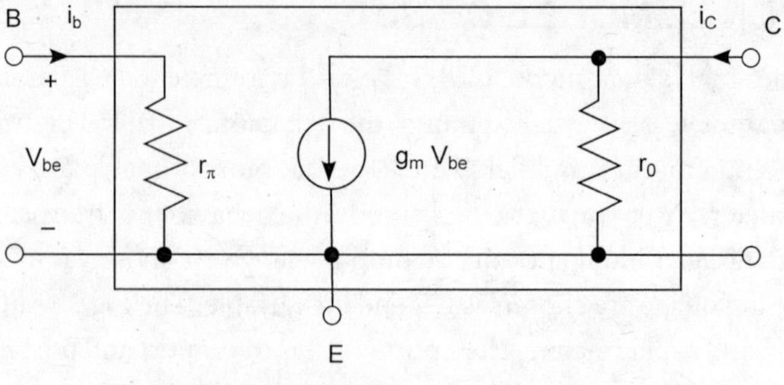

Fig. 2.10: Hybrid-π model of BJT

Here, g^m is a small signal parameter which relates small signal collector current to small signal base emitter voltage. Its expression is derived using (2.7) and is given in (2.19)

$$g_m = \frac{\partial I_C}{\partial V_{BE}} = \frac{I_C}{V_T} \tag{2.19}$$

The r_π is input impedance and its expressionis derived using (2.8) and is given in (2.20),

$$r_\pi = \left(\frac{\partial I_B}{\partial V_{BE}}\right)^{-1} = \left(\frac{1}{\beta_F}\frac{\partial I_C}{\partial V_{BE}}\right)^{-1} = \frac{\beta_F}{g_m} \tag{2.20}$$

Further, r_o is derived using (2.10) and given in (2.21),

$$r_o = \left(\frac{\partial I_C}{\partial V_{CE}}\right)^{-1} = (g_o)^{-1} = \frac{V_A}{I_C} \tag{2.21}$$

Small signal parameters given in (2.19), (2.20) and (2.21) depends upon bias points. Hence, biasing not only decides the region of BJT operation but also helps in fixing the value of small signal parameter. it is important to fix bias points based on desired small signal performance. Once bias points are fixed so the small signal parameter and hence the small signal performance to certain extent.

Now steps to analyze the performance of any circuit which is designed using BJT or any other nonlinear elements will be following:

1. Find the bias points or operating points by solving the DC equivalent circuit. For finding the DC equivalent circuit, one should disable the small signal sources present in the circuit. If there are small signal voltage sources then it is short circuited and if there are small signal current sources then it is open circuited. All coupling capacitors are made open as capacitor offer high impedance at DC. Now apply KVL or/and KCL to solve for currents and voltages in this DC equivalent circuit. These currents and voltages are operating points of this circuit. Use operating points to find small signal parameters.
2. Draw the small signal equivalent circuit using small signal parameters found in step 1. For finding small signal equivalent circuit, one can null all DC sources present in circuit by shorting DC voltage sources and making DC current sources open circuited. The coupling capacitors are shorted in this

case as it is selected to offer very low impedance at signal frequencies. All nonlinear devices are substituted by its small signal equivalent circuit while linear devices remains same in small signal circuit. This small signal circuit is now solved using KVL or/and KCL to find the small signal performance of circuit.

Example 2.1:

Find operating points of the circuit shown in Fig.2.11(a) when $I_S = 10\mu$ A and $\beta_F = 99$. Also evaluate its small signal parameters and draw its small signal equivalent circuit.

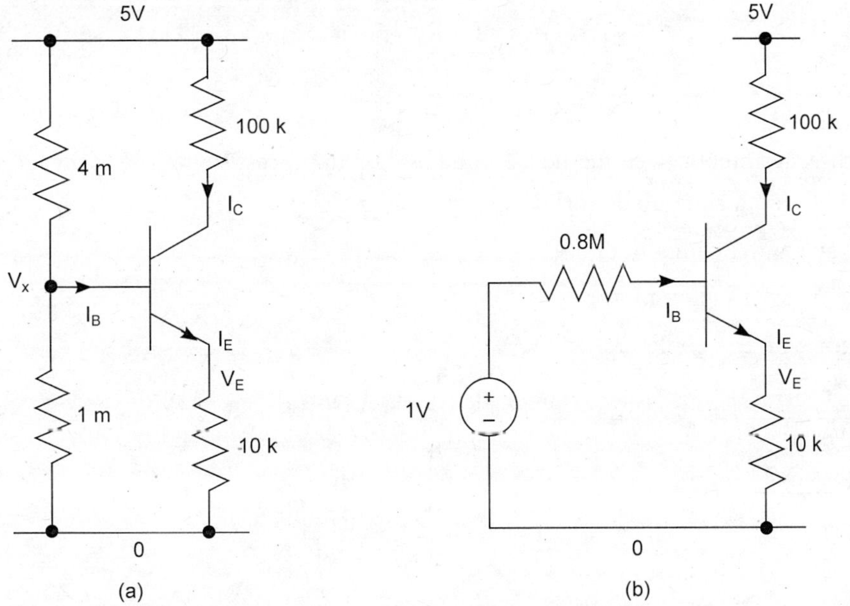

Fig. 2.11: Circuit for example 2.1

Solution:

One can simplify the circuit of Fig.2.11(a) into circuit of Fig.2.11(b) using Thevenin equivalent of resistor divider at the input of base terminal.

Now assuming that BJT is on and then solving for base current using KVL across base emitter loop we get (2.22)

$$1 - I_B \cdot 0.8\text{M} - V_{BE} - I_E \cdot 10\text{k} = 0 \tag{2.22}$$

On using (2.11) and (2.22) we get (2.23)

$$1 - I_B \cdot 0.8\text{M} - V_{BE} - (\beta_F + 1) \cdot I_B \cdot 10\text{k} = 0 \tag{2.23}$$

(2.23) is further simplified to solve for base current I_B as given in (2.24)

$$I_B = \frac{1 - V_{BE}}{0.8M + (\beta_F + 1) \cdot 10k} = \frac{1 - 0.7}{0.8M + (99 + 1) \cdot 10k} = 0.166\mu A \tag{2.24}$$

Now collector current is obtained using (2.3) as given in (2.25),

$$I_C = \beta_F I_B = 99 \times 0.166\ \mu A = 16.5\ \mu A \tag{2.25}$$

Further emitter current is obtained using (2.24) and (2.11) and is given in (2.26)

$$I_E = (1 + \beta_F)I_B = 100 \times 0.166\ \mu A = 16.6\ \mu A \tag{2.26}$$

Thus emitter voltage V_E is given using (2.27)

$$V_E = I_E \cdot R_E = 16.6\ \mu A \times 10\ k\Omega = 166\ mV \tag{2.27}$$

And collector voltage is calculated as given in (2.28),

$$V_C = V_{DD} - I_C \cdot R_C = 5 - (16.5\ \mu A) \times 100\ k\Omega = 3.35V \tag{2.28}$$

2.7 BJT CIRCUIT CONFIGURATION

As BJT is a three terminal device and hence while using it to process signal one of the terminal is made common to both input and output port. Hence there are three possible configuration of BJT circuits.

1. Common emitter configuration
2. Common base configuration
3. Common collector configuration

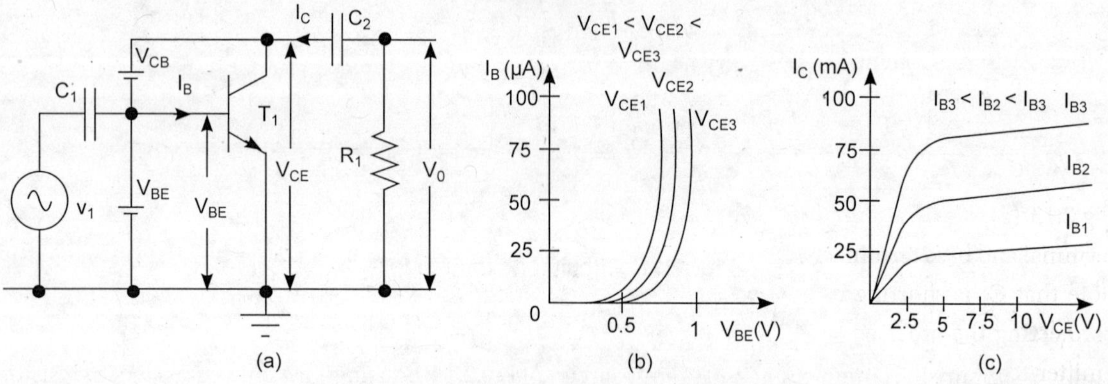

Fig. 2.12: (a)Common emitter (CE) configuration of BJT (b) Input characteristic of CE configuration (c) output characteristic of CE configuration

2.7.1 Common Emitter Configuration

Fig.2.12(a) shows a common emitter configuration of BJT. In this configuration, input signal is applied between base and emitter terminal of BJT and output is taken across collector and emitter terminal of BJT. As emitter terminal is common to both input port and output port, hence this configuration is known as common emitter configuration. In this configuration, input parameters are base current I_B and base-emitter voltage V_{BE} while output parameters are collector current I_C and collector-emitter voltage V_{CE}. The behaviour of amplifier configuration can also be described using its input and output characteristic. The input characteristic of CE configuration is essentially a plot of I_B as a function of V_{BE} for constant value of V_{CE} as shown in Fig.2.12 (b) while output characteristic of CE configuration is a plot of I_C as a function of V_{CE} for constant value of I_B as shown in Fig. 2.12 (c). The h parameter of CE configuration are given in (2.29),

$$h_{ie} = r_\pi, h_{re} = 0, h_{fe} = \beta_F, h_{oe} = \frac{1}{r_o} \tag{2.29}$$

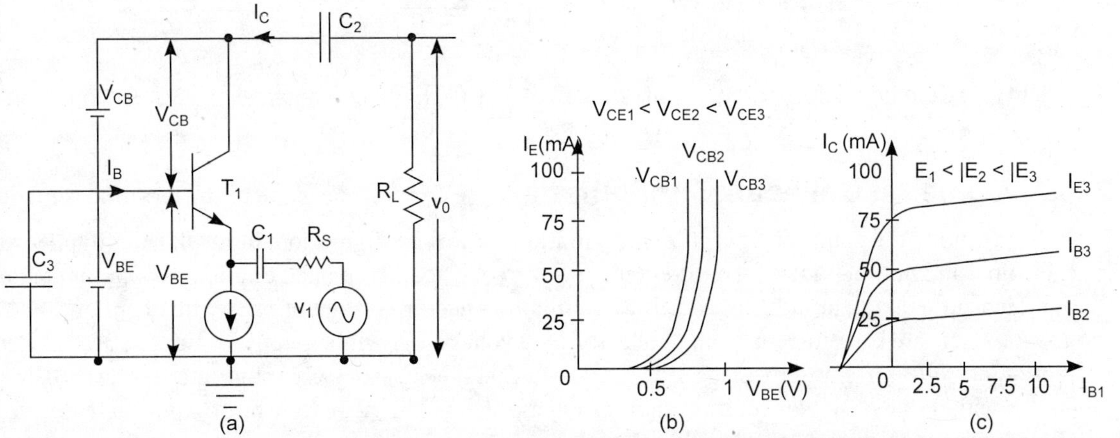

Fig. 2.13: (a) Common base (CB) configuration of BJT (b) Input characteristic of CB configuration (c) output characteristic of CB configuration

2.7.2 Common base Configuration

Fig.2.13 (a) shows a common base configuration. In this case, input signal is applied across emitter terminal and base terminal and output signal is taken across collector terminal and base terminal. Please note that C_1 is shorting one end of input source vi to emitter terminal of BJT while C_3 is effectively connecting the other end of vi to base terminal, Hence input source is connected between base and emitter terminal. Similarly load resistor is connected to collector terminal through C_2 at one end and to base through C_4 at other end, hence output is effectively taken across collector and base terminal of BJT. As base terminal is common to both input and output and hence this configuration is known as common base configuration. In this configuration, input parameters are emitter current I_E and base-emitter voltage V_{BE} while output parameters are collector current I_C and collector-base voltage V_{CB}. The input characteristic of CB configuration is essentially a plot of I_E as a function of V_{BE} for constant value of V_{CB} as shown in Fig.2.13 (b) while output characteristic of CB configuration is a plot of I_C as a function of V_{CB} for constant value of I_E as shown in Fig. 2.13 (c).

The *h* parameter of CB configuration are given in (2.30),

$$h_{ib} = \frac{1}{g_m}, h_{re} = 0, h_{fe} = \frac{\beta_F}{1+\beta_F}, h_{oe} = \frac{1}{R_S(1+g_m r_o)} \tag{2.30}$$

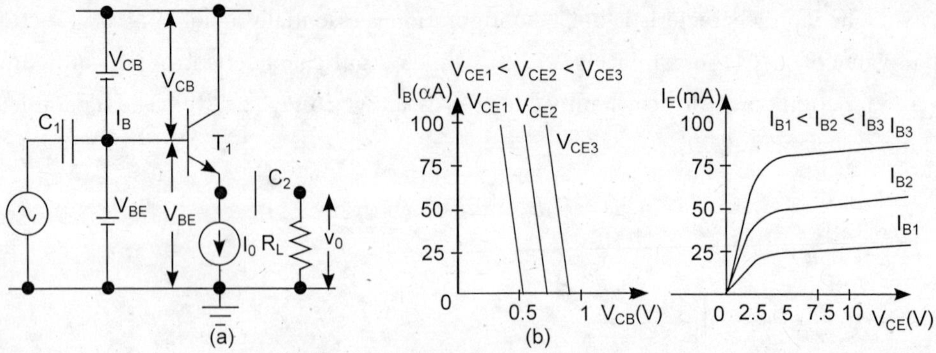

Fig. 2.14: (a) Common collector (CC) configuration of BJT (b) Input characteristic of CC configuration (c) output characteristic of CC configuration

2.7.3 Common Collector Configuration

Fig. 2.14 (a) shows the common collector configuration. In this configuration, input signal is applied at base terminal and output is taken at emitter terminal. As collector terminal is common to both input and output, Hence this configuration is known as common collector configuration. In this configuration, input parameters are base current I_B and collector-base voltage V_{CB} while output parameters are emitter current I_E and collector-emitter voltage V_{CE}. The input characteristic of CC configuration is essentially a plot of I_B as a function of V_{CB} for constant value of V_{CB} as shown in Fig. 2.14 (b) while output characteristic of CB configuration is a plot of I_C as a function of V_{CB} for constant value of I_E as shown in Fig. 2.14 (c). The h parameter of CC configuration are given in (2.31),

$$h_{ib} = r_\pi + (1+\beta_F)R_L, h_{re} = 0, h_{fe} = 1+\beta_F, h_{oe} = \frac{1}{\dfrac{1}{g_m} + \dfrac{r_\pi}{\beta_F}} \tag{2.31}$$

2.8 AMPLIFIER DESIGN USING BJT

One of the prominent application of BJT is to design amplifier. The details of different amplifier configurations designed using BJT are given below,

2.8.1 Amplifier using BJT in Common Emitter Configuration

The Fig. 2.15 (a) shows a common emitter configuration of the amplifier designed using BJT. The part of the circuit shown in blue is essentially the biasing circuit which ensures that BJT (T_1) is biased in active region of BJT operation. The values of components are chosen based on expected small signal parameters. The capacitors shown in red ensures that DC bias points of T_1 are not affected by the small signal source and load. As capacitors are open at DC, therefore, source and load is effectively disconnected from the transistor circuit. These capacitors are termed as coupling capacitor and its value is chosen in such a way that at signal frequencies these capacitors effectively get shorted. Fig. 2.15 (b) shows the small signal equivalent of circuit shown in Fig. 2.15 (a) where coupling capacitors $C_{1,2}$ are shorted as its impedance at signal frequencies are low. Now transistor can also be replaced with its small signal equivalent circuit which brings us to Fig. 2.15 (c). The circuit shown in Fig. 2.15 (c) is analyzed to find the small signal performance of this circuit.

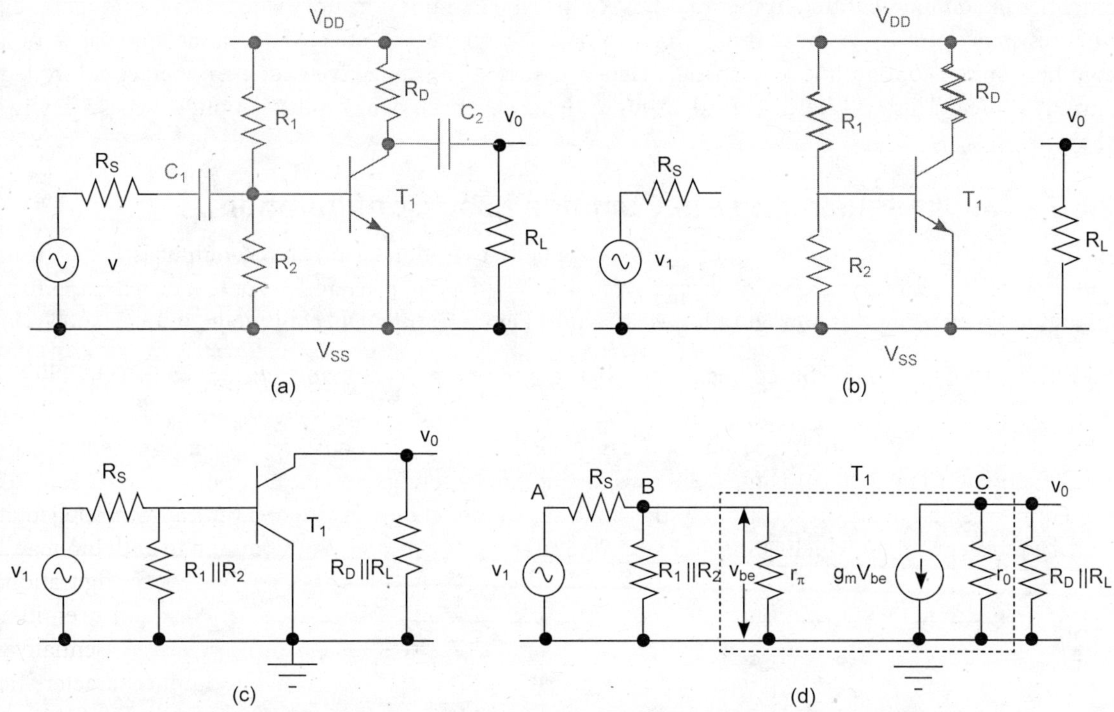

Fig. 2.15: (a) Common Emitter configuration (b) DC equivalent circuit (c) Small signal equivalent of circuit (d) Small signal equivalent circuit including small signal model of Transistor

The small signal gain of common emitter amplifier is obtained by solving its equivalent small signal circuit shown in Fig. 2.15 (c).

Node potential at node B which is v_{be} is obtained by solving the resistive divider at input side. It is given by (2.32)

$$v_{be} = \frac{(R_1 \,||\, R_2) + r_\pi}{R_S + (R_1 \,||\, R_2) + r_\pi} v_i \tag{2.32}$$

Now on solving the circuit at output node, v_o is given using (2.33)

$$v_o = -g_m \, v_{be} \, (R_D \,||R_L \,||r_o) \tag{2.33}$$

Using (2.32) and (2.33) we can obtain the amplifier gain as given in (2.34)

$$A_v = \frac{v_o}{v_i} = -g_m \frac{(R_1 \,||\, R_2) + r_\pi}{R_S + (R_1 \,||\, R_2) + r_\pi} (R_D \,||\, R_L \,||\, r_o) \tag{2.34}$$

If output impedance of input source is low then approximate amplifier gain is given using (2.35)

$$A_v = \frac{v_o}{v_i} = -g_m (R_D \,||\, R_L \,||\, r_o) \tag{2.35}$$

The impedance looking into base terminal of transistor which is effectively the input of amplifier when it is used in common emitter configuration is r_π which is given by (2.14) and is generally a high value. Hence the input impedance of a common emitter configuration is high. The output impedance of amplifier in common emitter configuration is $R_D \| r_o$ when amplifier is not loaded. Here, r_o is generally high resistance and R_D is chosen to be high to have a high gain so effectively output impedance of a common emitter configuration is also high. Hence in this case BJT effectively act like a Voltage controlled current source. This will be discussed in more detail when we will discuss amplifiers in successive chapters.

2.8.2 Amplifier using BJT in Common base Configuration

In common base configuration, input signal is applied at emitter terminal and output is taken from collector terminal of transistor while base terminal is biased using constant DC potential. The amplifier circuit designed using this configuration is shown in Fig.2.16(a). Its DC equivalent circuit is shown in Fig. 2.16 (b). Here current source I_o is used to bias the transistor T_1. It fixes collector current to $\frac{\beta_F I_o}{1+\beta_F}$ which is almost constant for large value of β_F and thus this biasing is robust. The small signal equivalent circuit shown in Fig. 2.16 (c) is obtained by shorting the coupling cap and nullifying the DC sources present in circuit. Finally on substituting the small signal model of transistor we get the complete small signal equivalent circuit which is shown in Fig.2.16 (d).

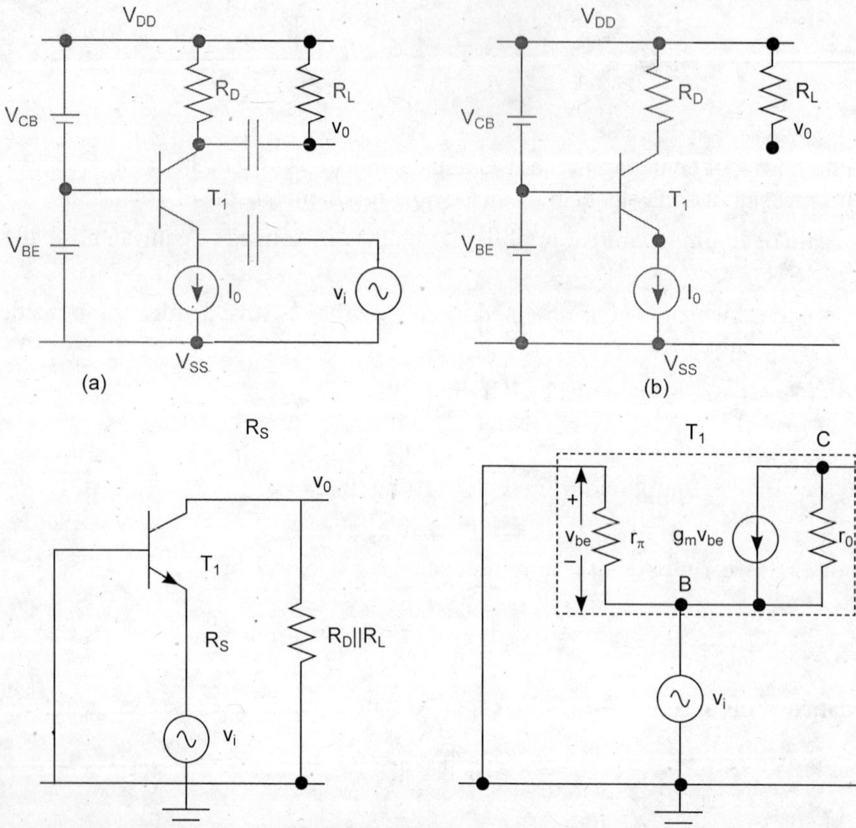

Fig. 2.16: (a) Common base configuration (b) DC equivalent circuit (c) Small signal equivalent of circuit (d) Small signal equivalent circuit including small signal model of Transistor

Now to calculate the small signal gain of this circuit one can write KVL for the left loop of the circuit shown in Fig.2.16(d) which is given in (2.36)

$$v_{be} + v_i = 0 \tag{2.36}$$

Now writing KCL at output node of the circuit shown in Fig. 2.16 (d) which is given by (2.37)

$$g_m v_{be} + \frac{v_o + v_{be}}{r_o} + \frac{v_o}{R_D \| R_L} = 0 \tag{2.37}$$

On simplifying (2.37) we obtain (2.38)

$$v_{be} = \frac{-v_o r_o}{(R_D \| R_L \| r_o)(1 + g_m r_o)} \tag{2.38}$$

Using (2.37) and (2.38) we obtain (2.39)

$$\frac{v_o}{v_i} = (g_m + \frac{1}{r_o})(R_D \| R_L \| r_o) \approx (g_m)(R_D \| R_L \| r_o) \tag{2.39}$$

2.8.3 Amplifier Design using Common Collector Configuration

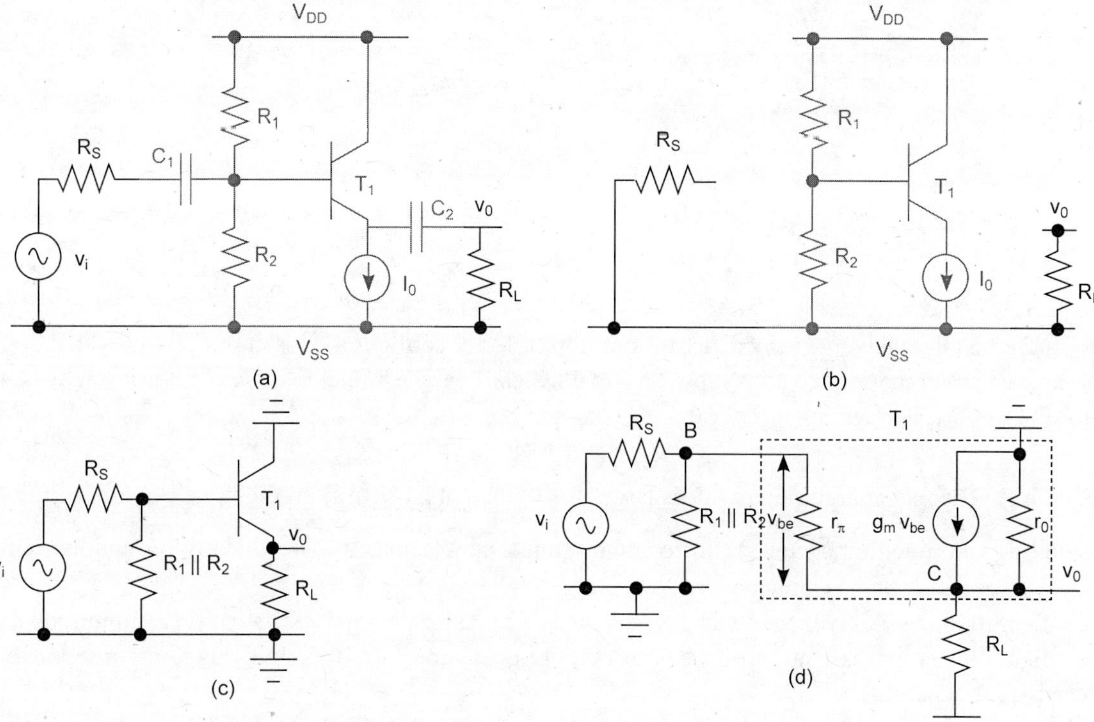

Fig. 2.17: (a) Common collector configuration (b) DC equivalent circuit (c) Small signal equivalent of circuit (d) Small signal equivalent circuit including small signal model of Transistor

In common collector configuration, input signal is applied at base terminal and output is taken from emitter terminal of transistor while collector terminal is biased using constant DC potential. The amplifier circuit designed using this configuration is shown in Fig. 2.17(a). Its DC equivalent circuit is shown in Fig. 2.17 (b). Here current source I_o is used to bias the transistor T_1. It fixes collector current to $\frac{\beta_F I_o}{1+\beta_F}$ which is almost constant for large value of β_F and thus this biasing is robust. The small signal equivalent circuit shown in Fig. 2.17 (c) is obtained by shorting the coupling cap and nullifying the DC sources present in circuit. Finally on substituting the small signal model of transistor we get the complete small signal equivalent circuit which is shown in Fig.2.17(d).

Now to calculate the small signal gain of this circuit one can first find potential at node B using voltage divider as given in (2.40)

$$v_B = \frac{R_1 \| R_2}{R_S + R_1 \| R_2} v_i \approx v_i \tag{2.40}$$

Now writing KCL at output node of the circuit shown in Fig. 2.17 (d) as given by (2.41)

$$\frac{v_B - v_o}{r_\pi} + g_m (v_B - v_o) = \frac{v_o}{R_L \| r_o} \tag{2.41}$$

On simplifying (2.41) we obtain (2.42)

$$v_o = \frac{v_B (R_L \| r_o \| r_\pi \| \frac{1}{g_m})}{(r_\pi \| \frac{1}{g_m})} \approx v_B \tag{2.42}$$

Using (2.40) and (2.42) we obtain (2.43)

$$\frac{v_o}{v_i} \approx 1 \tag{2.43}$$

The small signal input impedance in common collector configuration is the current provided by test source when it is connected at its input terminal which is base terminal in this case. It is given by (2.44) for the amplifier shown in Fig.2.17(a)

$$Z_{in} = R_S + ((R_1 \| R_2) \| r_\pi + (\beta_F + 1) R_L) \tag{2.44}$$

It is generally a large value as β_F is large and R_L is kept large to keep the gain close to unity. The small signal output impedance of this configuration is approximately $\frac{1}{g_m}$ which is low and hence this configuration acts like a voltage controlled voltage source. As small signal voltage gain of common collector configuration is approximately unity and its output impedance is low therefore this configuration is also used as a voltage buffer.

2.9 HIGH FREQUENCY MODEL OF BJT

The hybrid-π model of BJT discussed in Fig.2.10 does not model the frequency dependent behaviour of BJT. However, like any other real component, BJT do have some frequency dependent behaviour due to parasitic capacitances present in it. Hence, the analysis done using this model deviate significantly from the real performance at high frequency. Therefore, a better model is needed to predict thefrequency dependent behaviour of BJT. The high frequency behaviour of BJT is effectively modelled by adding two more capacitor in hybrid-π model as shown in Fig.2.18. The capacitor C_π is a combination of two effects, one is due to base charge denoted as C_b while other is due to junction capacitance at base-emitter junction denoted as C_{je}. Hence $C_\pi = C_b + C_{je}$. Further, C_μ accounts for junction capacitance at base-collector junction

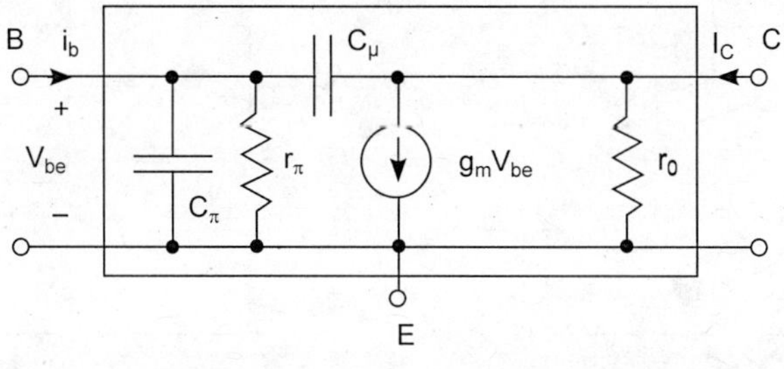

Fig. 2.18: High frequency small signal model of BJT

Now due to these capacitance, current gain of transistor h_{fe} falls at high frequencies. The frequency at which current gain of transistor becomes unity is defined as transit frequency f_T. Thus, to find the transit frequency one can first find current gain and equate it to unity. To calculate current gain, one need to short circuit the output as suggested in (2.17). The circuit becomes as the one shown in Fig.2.19 after doing it. In this circuit r_o does not play any role as it is effectively shorted and also v_{be} appears across C_μ as well as other end of it is shorted to emitter.

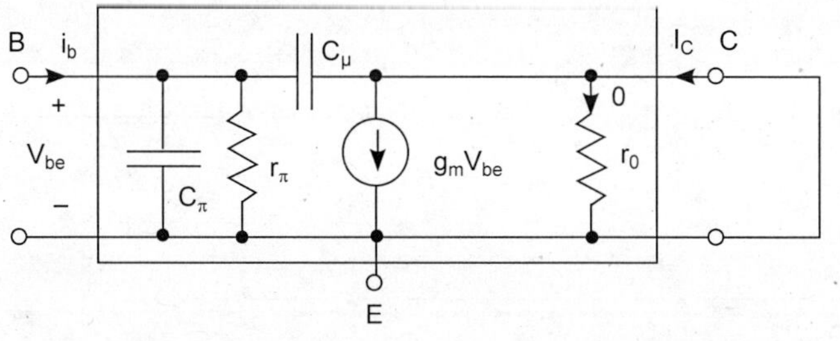

Fig. 2.19: Circuit used to calculate transit frequency

Now at high frequency impedance offered by r_π becomes much larger than capacitances and hence base current will flow only through capacitances. Thus v_{be} is given using (2.45)

$$v_{be} = i_b\, s(C_\pi + C_\mu) \tag{2.45}$$

Further output current i_C is given using (2.46)

$$i_c = g_m\, v_{be} - sC_\mu\, i_b \approx g_m\, v_{be} \tag{2.46}$$

Hence using (2.45) and (2.46) we can find the current gain expression given in (2.47)

$$\frac{i_c}{i_b} = \frac{g_m}{s(C_\pi + C_\mu)} \tag{2.47}$$

Now, transit frequency f_T by obtained by equating (2.47) to unity, which gives (2.48),

$$s = \frac{g_m}{C_\pi + C_\mu} \Rightarrow f_T = \frac{g_m}{2\pi(C_\pi + C_\mu)} \tag{2.48}$$

Example: 2.2:

For the circuit shown in Fig. 2.20(a) Draw its high frequency small signal model.

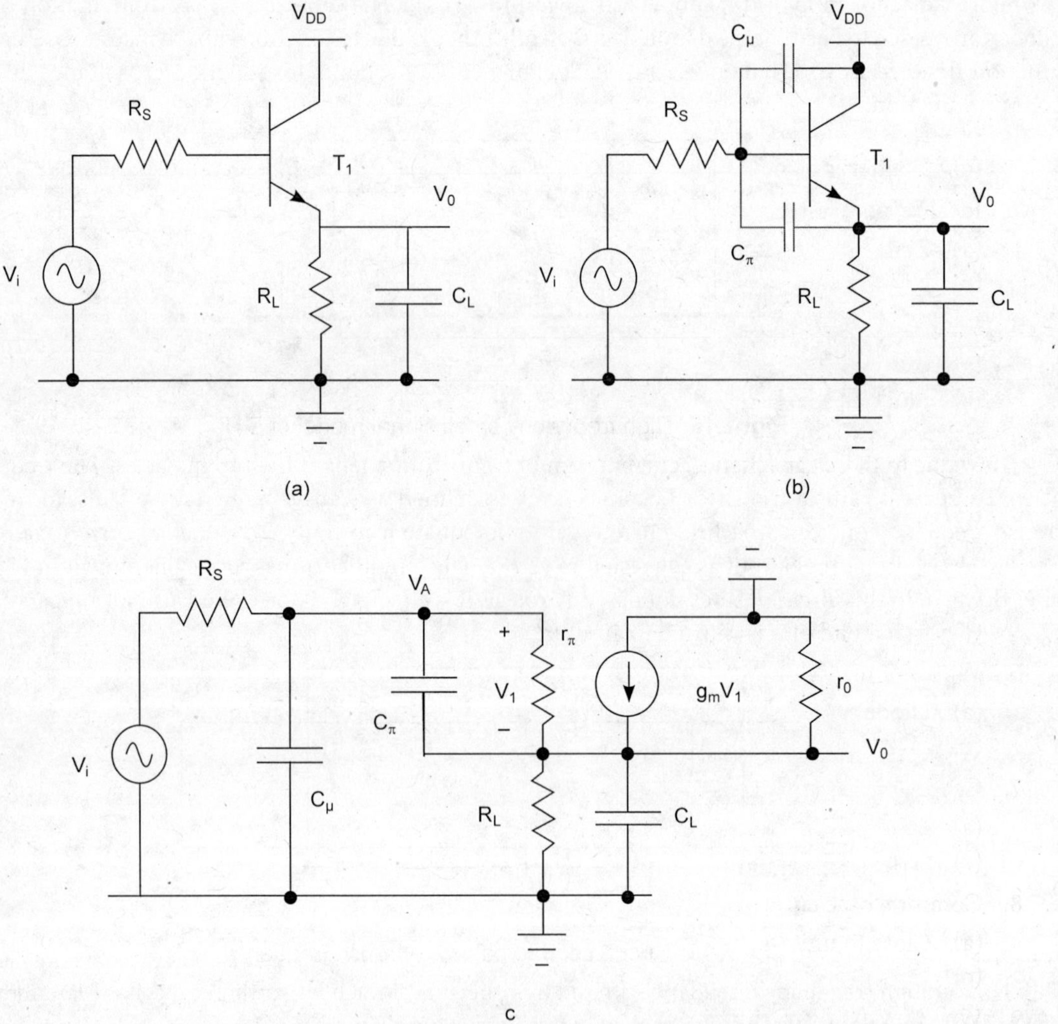

Fig. 2.20: (a) Example circuit of CC configuration (b) Circuit with parasitic capacitances of BJT (c) High frequency small signal equivalent circuit of example circuit in part(a)

Solution:

The circuit shown in Fig. 2.20 (a) is a common collector amplifier which is also known as emitter follower. The Fig. 2.20 (b) shows the circuit with parasitic capacitance of T_1. Its small signal equivalent circuit is shown in Fig. 2.20 (c).

UNIT SUMMARY

In this unit, basic functioning and solid state physics behind BJT has been introduced. It develops small signal DC and AC models of BJT operation. Several different BJT configurations are discussed and basic amplifiers are designed using these configurations.

EXERCISES

Multiple Choice Questions

1. Which of the following statement is correct?
   (a) Base region is heavily doped in BJT.
   (b) Collector is heavily doped in BJT.
   (c) Emitter is heavily doped in BJT.
   (d) Base is much thicker than other two regions.
2. Ideal reverse voltage gain of BJT is __ ? Fill in the blanks with correct option.
   (a) Zero
   (b) Infinity
   (c) Negative
   (d) finite positive number
3. BJT is a __ device. Fill in the blanks with one of the following options.
   (a) Unipolar
   (b) Bipolar
   (c) Junction less
   (d) Passive
4. Which of the following region is highly doped in BJT
   (a) Base
   (b) Collector
   (c) Emitter
   (d) None of the above
5. Which of the following region has smallest length in BJT?
   (a) Base
   (b) Emitter
   (c) Collector
   (d) None of the above
6. Which of the following device is a current controlled device
   (a) Diode
   (b) Resistor
   (c) BJT
   (d) Transformer
7. Common emitter current gain β_F of BJT when it operates in active region is
   (a) Low
   (b) High
   (c) zero
   (d) Close to unity
8. Common base current gain α_F of BJT when it operates in active region is
   (a) Close to unity but more than one
   (b) Close to unity but less than one
   (c) zero
   (d) Very High
9. Which of the following bias condition defines active region of BJT operation
   (a) E-B junction is forward biased and B-C junction is reverse biased
   (b) Both E-B and B-C junctions are forward biased
   (c) Both E-B and B-C junctions are reverse biased
   (d) E-B junction is reverse biased and B-C junction is forward biased
10. Which of the following bias condition defines saturation region of BJT operation
    (a) E-B junction is forward biased and B-C junction is reverse biased
    (b) E-B junction is reverse biased and B-C junction is forward biased
    (c) Both E-B and B-C junctions are forward biased
    (d) Both E-B and B-C junctions are reverse biased
11. Which of the following are majority charge carriers in n-p-n transistor
    (a) Holes
    (b) Electrons
    (c) Neutrons
    (d) Both (a) and (b)

12. BJT is operated in which region of operation to act as an amplifier

(a) Saturation (b) Active

(c) Cutoff (d) Reverse Active

13. Which of the following region of operations of BJT are used to perform switching action

(a) Cutoff and Reverse active (b) Cutoff and active

(c) Active and saturation (d) Cutoff and saturation

14. Which of the following is correct definition of transit frequency of a BJT transistor

(a) The frequency at which voltage gain of BJT becomes unity

(b) The frequency at which voltage gain of BJT is 3dB less than its DC voltage gain

(c) The frequency at which current gain of BJT becomes unity

(d) The frequency at which current gain of BJT is 3dB less than its DC current gain

15. Voltage gain of common collector amplifier designed using BJT is

(a) Zero (b) very high

(c) Close to unity but less than one (d) Close to unity but more than one

16. Which of the following amplifier configuration offers least output impedance

(a) Common emitter configuration (b) Common base configuration

(c) Common collector configuration

17. Which of the following amplifier configuration offer highest input impedance

(a) Common emitter configuration (b) Common base configuration

(c) Common collector configuration

ANSWERS															
1.	(c)	2.	(a)	3.	(b)	4.	(c)	5.	(a)	6.	(c)	7.	(b)	8.	(b)
9.	(a)	10.	(c)	11.	(b)	12.	(b)	13.	(d)	14.	(c)	15.	(c)	16.	(c)
17.	(c)														

Short Answer Type Questions

1. Why two p-n junction diode available in laboratory when connected back to back does not function like a BJT device?
2. Why BJT is defined as a bipolar device?
3. Why BJT is defined as a current controlled device?
4. What is common emitter current gain and common base current gain? How these two parameters are related?
5. What is the difference between diode and transistor?
6. Which of the BJT configuration has highest input impedance?
7. What is transit frequency? Give the expression of it?
8. Provide the expression of transconductance and output resistance of BJT transistor.

Long Answer Type Questions

1. describe the different region of operation of a BJT transistor?
2. What is hybrid-π model of a BJT transistor? Derive the expression of small signal parameters used in it.
3. Compare the different configuration of BJT transistor. Point out the advantages and disadvantages of each of these configurations.
4. Derive the expression of input impedance of common collector configuration shown in Fig.2.17(a).
5. Derive the expression of transit frequency of a BJT transistor.

Numerical Problems

1. A BJT transistor is biased such that its collector current is 1mA. Find its transconductance.
2. A BJT has a base current of 0.1mA and emitter current of 1mA. Find its collector current. What is its common emitter current gain.
3. The early voltage of a BJT transistor is given as 100V. If its collector current is 1mA then find its output resistance.
4. If common base current gain of a BJT is 0.99 then find its common emitter current gain.
5. Find input base resistance r_π of BJT if its common emitter current gain is 100 and its collector current is 1mA.
6. A BJT has a common emitter current gain equal to 100. If its base current is 0.1mA then find its collector current. Also find its transconductance. If its early voltage is 100V then find its output resistance.
7. For the circuit shown in Fig.2.21, V_{BE} of BJT is considered as 0.7V. Find its small signal parameters. Consider common emitter current gain β_F of BJT as 99 and thermal voltage V_T = 40mV.Find small signal gain v_o/v_i, input impedance R_i ,output impedance R_o. Assume early voltage of BJT is 100V.

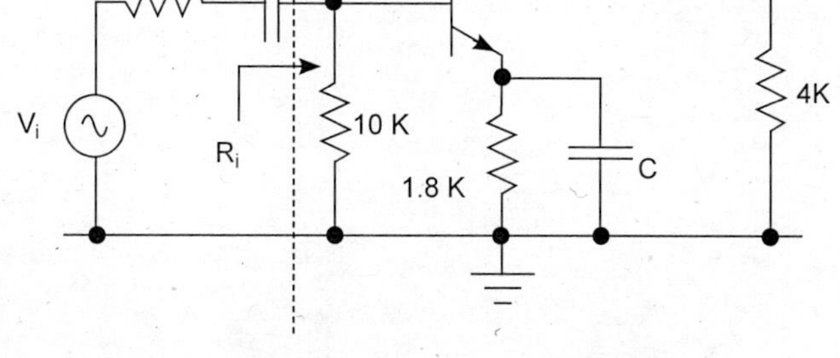

Fig. 2.21: Circuit for Q.N2.7

KNOW MORE

It has been around 75 years since the transistor has been demonstrated in its working form. The transistor is indeed one of the most significant invention in the field of semiconductor design. In fact it has revolutionized all aspects of life. Recently several noted authors have attempted to summarize the history and contribution of transistor through various review and research articles. One such link is mentioned here to get the more details about its history.

REFERENCES AND SUGGESTED READINGS

1. Bardeen, John, and Walter Houser Brattain. "Physical principles involved in transistor action." *Physical Review* 75, no. 8 (1949): 1208.
2. Riordan, Michael, Lillian Hoddeson, and Conyers Herring. "The invention of the transistor." *Reviews of Modern Physics* 71, no. 2 (1999): S336.
3. Shockley, William. "Transistor physics." *American Scientist* 42, no. 1 (1954): 41-72.
4. Nelson, Richard. "The link between science and invention: The case of the transistor." In *The rate and direction of inventive activity: Economic and social factors*, pp. 549-584. Princeton University Press, 1962.
5. Shockley, William. "Transistor technology evokes new physics." *Nobel lecture* 11 (1956).
6. Shockley, William. "The path to the conception of the junction transistor." *IEEE Transactions on Electron Devices* 23, no. 7 (1976): 597-620.
7. Shockley, William. "The path to the conception of the junction transistor." *IEEE Transactions on Electron Devices* 31, no. 11 (1984): 1523-1546.

❑❑

3 Field Effect Transistor (FET)

UNIT SPECIFICS

Through this unit we have discussed the following aspects:

- Introduction to semiconductor physics of FET and its basic operation
- Introduction to Junction FET and its operation
- Introduction to Metal oxide semiconductor (MOS) FET and its basic operation
- Application of MOS devices, switch, amplifiers
- Introduction to unijunction transistors and its basic operation

The practical applications of the topics are discussed for generating further curiosity and creativity as well as improving problem solving capacity.

Besides giving a large number of multiple choice questions as well as questions of short and long answer types marked in two categories following lower and higher order of Bloom's taxonomy, assignments through a number of numerical problems, a list of references and suggested readings are given in the unit so that one can go through them for practice. It is important to note that for getting more information on various topics of interest some QR codes have been provided in different sections which can be scanned for relevant supportive knowledge.

After the related practical, based on the content, there is a "Know More" section. This section has been carefully designed so that the supplementary information provided in this part becomes beneficial for the users of the book. This section mainly highlights the initial activity, examples of some interesting facts, analogy, history of the development of the subject focusing the salient observations and finding, timelines starting from the development of the concerned topics up to the recent time, applications of the subject matter for our day-to-day real life or/and industrial applications on variety of aspects, case study related to environmental, sustainability, social and ethical issues whichever applicable, and finally inquisitiveness and curiosity topics of the unit.

RATIONALE

This fundamental unit on field effect transistor (FET) helps students to get a primary idea about the basic operation as well as semiconductor physics behind the operation of FET. It covers different types of FET providing details about its operation and circuit design. Semiconductor physics behind Metal oxide semiconductor (MOS) FET is first explained using basic structure of MOS capacitor and then extended to MOS transistors. It also discusses the method of performing small signal analysis on circuits designed using MOSFET transistors.

Further, different amplifier configurations are discussed. Description of unijunction transistor is provided to complete the discussion on FET.

PRE-REQUISITES

Mathematics: Calculus (Class XII)

Physics: Semiconductor physics (Class XII)

UNIT OUTCOMES

List of outcomes of this unit is as follows:

U3-O1: Introduction and classification of FET

U3-O2: Description of working principle of JFET

U3-O3: Description of working principle of MOS capacitor, MOSFET

U3-O4: Analysis of circuits designed using MOSFET

U3-O5: Introduction to Unijunction transistor (UJT)

Unit-3 Outcome	EXPECTED MAPPING WITH COURSE OUTCOMES (1 - Weak Correlation; 2 - Medium Correlation; 3 - Strong Correlation)				
	CO-1	CO-2	CO-3	CO-4	CO-5
U3-O1	3	1	2	1	1
U3-O2	3	2	2	1	1
U3-O3	3	2	2	1	2
U3-O4	2	3	3	1	3
U3-O5	3	3	2	1	1

3.1 INTRODUCTION

In this chapter we study a class of transistors where current flowing through transistor is controlled by applying a voltage at its control terminal. This applied voltage effectively creates an electric field which influences charge flow between its two other terminals. As current is controlled by application of an electric field hence this category of transistors are commonly known as filed effect transistor (FET). Unlike BJT, FETs are unipolar device which means current conduction in FET is due to one type of charge carrier, either holes or electrons.

The concept of FET was first patented by physicist Julius Edger Lilienfeld in 1925 but he was unable to build a working prototype of this device. This was due to problem with surface states and dangling bonds in materials being used for it such as germanium and copper. Later on, first FET device successfully fabricated in 1952 was a variant of Junction Field effect transistor (JFET) which was patented by Heinrich Welker in 1945. However, design and fabrication of FET still was an issue due to its large size which restricted its mass production. Later on the work envisioned by William Shockley, John Bardeen and Walter Brattain paved the way for the design and fabrication of another variant of FET known as metal oxide semiconductor field effect transistor (MOSFET) by Mohamed Atalla and DawonKahng at Bell labs in 1959. The MOSFET has revolutionized the semiconductor industry and led for the development of integrated circuit (IC) design.

FETs are classified in two categories

1. Junction Field effect transistors (JFET)
2. Insulated gate field effect transistor (IGFET)

3.2 Junction Field Effect Transistor (JFET)

JFET was conceptualized much earlier than IGFET. Its input resistance is significantly higher than BJT which is helpful in reducing the power dissipation of digital circuits realized using JFET. But IGFET has even higher input resistance than JFET and can also provide many other benefits such as easy integration and scaling. This has made JFET almost obsolete and restricted its use as discrete devices built using JFET. JFETs are further classified in two categories, p-type JFET or n-type JFET depending upon the type of semiconductor material used to create its channel.

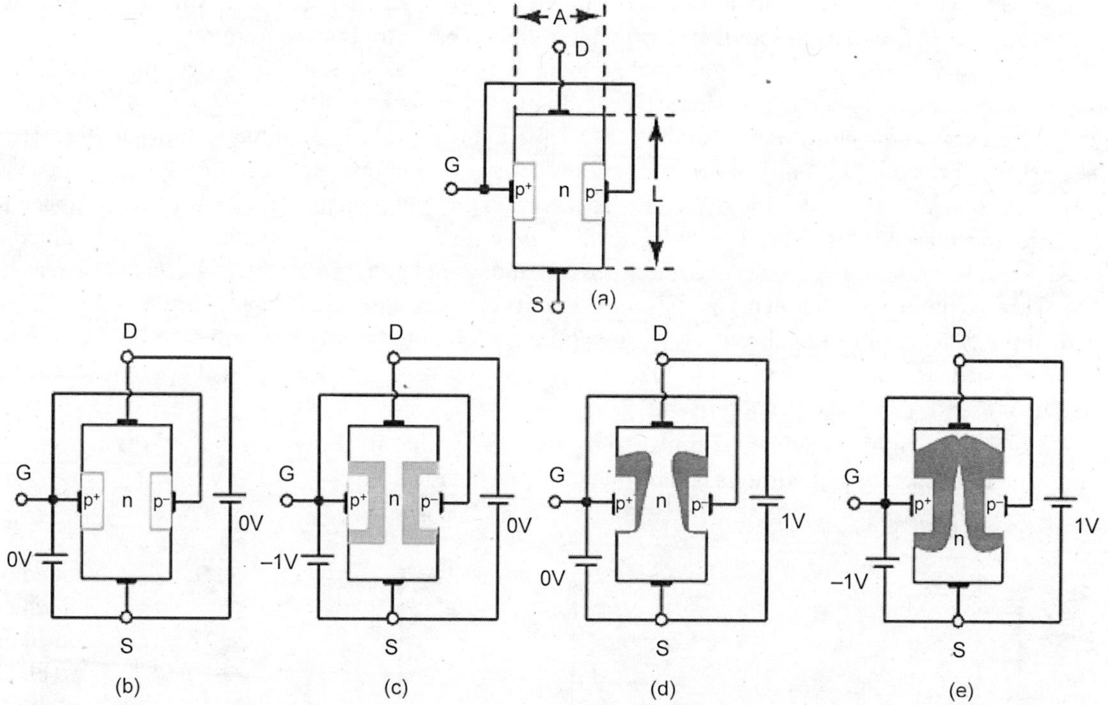

Fig. 3.1: (a) n-type JFET with V_{DS} = 0V and V_{GS} = 0V (b) n-type JFET with V_{DS} = 0V and V_{GS} =–1V (c) n-type JFET with VDS = 1V and V_{GS} = 0V (d) (a) n-type JFET with V_{DS} = 1V and V_{GS} = –1V

Fig.3.1(a) shows a representative diagram of a JFET. Here length (L) of device is generally much longer than its width (A). This device is essentially a slab of semiconductor material which in this case is of n-type and hence this device is known as n-type JFET. It has pockets of p^+-type diffusion along its sides. It also has contacts on top and bottom ends of it. These contact terminals are named as source (S) and drain (D). These terminals are named so because when a potential difference is applied between drain and source then source provides charge carriers, which are electrons in this case and these electrons are drained out by other terminal known as drain terminal. The source and drain terminal are interchangeable as device is completely symmetrical. In case of n-type JFET, terminal having higher potential among S and D acts as a drain terminal and other becomes source terminal. Further, one contact each are also provided on p⁺-type semiconductor material diffused along two sides of n-type channel. These two contacts are shorted together to form a Gate terminal (G).

Now, in order to understand the operation of this device, let us first consider S, D and G terminal are at same potential as shown in Fig.3.1(b). In this case p-n junctions are biased with zero potential where depletion region is relatively thin and mostly extended into n-type region as p-type region is heavily dopped. The depletion region is uniform along the p+-n junction. The part of n-type region which is not depleted is termed as channel as it can allow current to flows through it. In this case channel is widely open but no current flows through it as there is no potential difference across channel. Now if potential at G reduces, p^+-n junction becomes reverse biased causing depletion region to extend further in n-type semiconductor as shown in Fig.3.1(c). This reduces opening of channel and increases its resistance. However current flowing from D to S is still zero as potential difference across channel is zero. The potential V_{GS} at which channel opening becomes zero which means channel becomes void of any free charge is known as pinch off potential (V_P). So far, we have maintained V_{DS} as zero. Now, if V_{DS} increases then electric field gets established in channel which allows charge to flow from source to drain resulting in a current. This current creates a potential gradient in channel as channel is resistive and establishes a higher potential at part of channel close to drain than at part of channel close to source. This creates a wider depletion region close to drain and thinner depletion region close to source as shown in Fig.3.1(d). Thus channel opening becomes taper and is less at drain side compared to source side. Now as V_{DS} increases further the current through device also increases making p^+-n junction close to drain more and more reverse biased. This reduces channel opening even further. At some V_{DS} channel gets completely close and current through channel becomes constant. It does not increase any further even on increasing V_{DS}. This phenomenon is known as pinch-off. It occurs at $V_{DS} = V_{GS} - V_P$. Here V_P is pinch-off voltage. Now if potential at G terminal reduces as shown in Fig.3.1(e) then also p^+-n junction becomes more reverse biased causing depletion region to become even wider and thus increasing effective resistance of the channel and reducing current for same V_{DS}.

The working of p-type JFET is similar to n-type JFET with only difference that charge carriers are holes and thus operating conditions will be dual of this device.

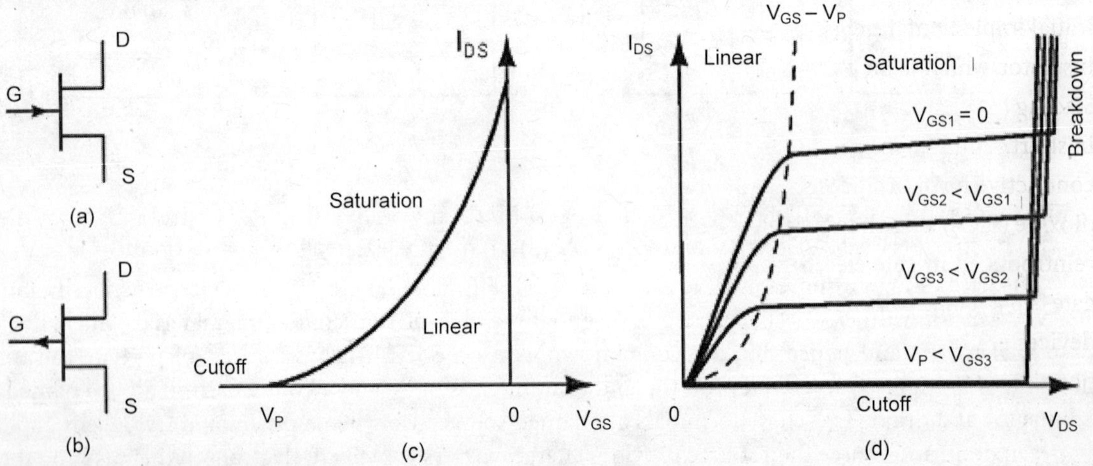

Fig. 3.2: (a) Symbol of n-type JFET (b) Symbol of p-type JFET (c) Transfer characteristic of n-type JFET (d) Output characteristic of n-type JFET

Fig.3.2(a) and (b) shows the symbol of n-type JFET and p-type JFET respectively. As p^+-n junction is always kept in reverse bias so gate current I_{GS} is very small in JFET and can be assumed zero for all practical purpose. The transfer characteristic of n-type JFET is shown in Fig.3.2(c). The equation of current I_{DS} is expressed using (3.1)

$$I_{DS} = I_{DSS}\left(1 - \frac{V_{GS}}{V_P}\right)^2 \tag{3.1}$$

Where, I_{DSS} is current flowing through device at $V_{GS} = 0$ and known as saturation current, V_{GS} is the potential difference between G and S terminal of JFET and V_P is pinch-off voltage of the JFET.

The output characteristic of JFET is shown in Fig.3.2 (d). The current I_{DS} is zero when V_{DS} is zero. Now, as V_{DS} increases, I_{DS} also increases but it increases almost linearly in the beginning. This region of operation of JFET is known as linear region as output characteristic is linear in this case. Now if V_{DS} increases further then current becomes saturated as depletion region due to reverse biased p^+-n region start modulating the resistance of the channel. Finally, I_{DS} becomes almost constant when V_{DS} reaches to $V_{GS} - V_P$ and does not increase much on increasing V_{DS}. This region of operation is known as saturation region because current has almost become saturated to a fixed value. However, if V_{DS} is increased beyond a point then avalanche breakdown happens across p+-n junction resulting in a flow of significantly large current. This region of operation is termed as breakdown region and device may get damaged if operated in this region.

3.3 METAL OXIDE SEMICONDUCTOR FIELD EFFECT TRANSISTOR (MOSFET)

Another variant of FET is Metal oxide semiconductor Field effect transistor (MOSFET) also known as MOS. Like JFET, MOS also uses a gate terminal to control the current flowing through it. In its initial implementation gate terminal is designed using metal and is separated from semiconductor by an insulator which is an oxide and hence the interfacing layers become Metal, Oxide and Semiconductor leading to the name of device as Metal Oxide Semiconductor Field effect transistor or MOSFET or MOS in short. Though, later on highly doped polycrystalline silicon also known as polysilicon which is highly conductive form of silicon is being used as gate material instead of metal as it does not react with silicon dioxide (SiO_2) at high temperature but the device is still called as MOS. After 2008, metal has again been reintroduced in gate electrode and oxide has been replaced with high-k dielectric material to increase gate capacitance without reducing dimensions in advanced node process. The other name used for these devices are Insulated gate field effect transistor (IGFET) which is more generic however MOS is more popular.

The use of JFET is now largely limited as discrete devices in low noise applications as its noise performance is better compared to MOS transistors. However, JFET is not scalable like MOS. In fact, MOS transistors have number of benefits in terms of size, power consumption and manufacturing yield when compared to JFET which make them preferable for integrated circuit (IC) design.

3.3.1 MOS capacitor

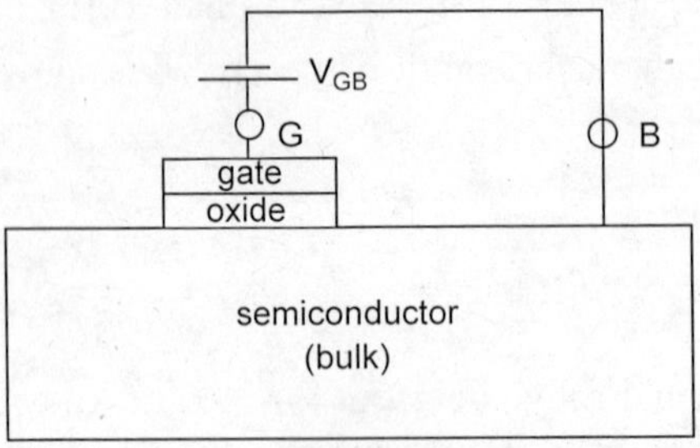

Fig. 3.3: MOS capacitor

To understand the operation of a MOS device, we first consider a much simpler device shown in Fig. 3.3 which is known as MOS capacitor. In order to fabricate this device, we start with a slice of semiconductor material which is sometime also referred as bulk material. The thickness of this slice is in the range of few tens of mm to several hundreds of mm. This slice is also known as wafer. Now, a very thin layer of oxide having a thickness of 1-2nm is grown on top of this semiconductor material. If semiconductor material chosen is silicon(Si) then oxide layer is generally silicon dioxide (SiO_2). Now, a highly conductive layer is grown on top of this oxide layer which forms the gate electrode. The contacts are provided to gate (G) and bulk (B) making it a two terminal device. This two terminal device is essentially a parallel plate capacitor where one of the plate of this capacitor is gate electrode and other plate is semiconductor or bulk. Now if both plates are assumed to be made up of same material and are shorted using the wire of same material then no potential difference is experienced while moving from one plate to another through external wire. Thus there will not be any build-up of charge on both sides of oxide layer. It is also assumed here that there is no charge build-up inside oxide layer during its formation. Though, in reality both of these assumptions are not true. Now, if gate electrode is made up of different material and not semiconductor then even on shorting the two electrode we will encounter several contact potentials in moving from gate electrode to semiconductor through external short. The net potential difference depends on the difference of contact potential of first material and last material. Further, some charges are trapped in the oxide layer or are at its interface depending upon its fabrication methods. In fact, for long these surface charges have limited the fabrication of MOS devices. However, now procedures are known which can minimize these charges. The net potential difference on both sides of oxide layer depends on difference of contact potential of gate material, bulk material and charges in oxides when external applied potential to device is zero. This potential results in build-up of charges on both sides of oxide layer. The external potential V_{GB} needed to balance out this charge to keep the device neutral everywhere is known as flat-band voltage of the device. Its expression is given using (3.2)

$$V_{FB} = \phi_G - \phi_B - \frac{Q_o}{C_{ox}} = \psi_{MS} - \frac{Q_o}{C_{ox}} \tag{3.2}$$

Where ϕ_G is contact potential of gate material, ϕ_B is contact potential of bulk material, Q_o Is total charge within oxide per unit area, C_{ox} is oxide capacitance per unit area and ψ_{MS} is the difference of contact potential between gate and semiconductor.

Another important point to mention here is that the whole system should maintain charge neutrality. Now, If effective charge on gate electrode is Q_G and effective charge in semiconductor under oxide layer which is also known as surface layer is Q_C then (3.3) must hold to maintain charge neutrality of the system.

$$Q_G + Q_0 + Q_C = 0 \tag{3.3}$$

In MOS Q_0 is assumed to be fixed and much smaller and hence are neglected in further analysis. Thus to keep charge neutrality any change in charge at gate electrode must be balanced out by corresponding change in charge in surface layer.

When external potential V_{GB} is applied then a part of it ψ_{ox} appears across oxide and another part ψ_s appears across semiconductor. Then going around the loop we can write (3.4).

$$V_{GB} = \psi_{ox} + \psi_s + \psi_{MS} = \psi_{ox} + \psi_s + V_{FB} \tag{3.4}$$

Now if V_{GB} is equal to V_{FB} then no potential is dropped across oxide and semiconductor and hence surface potential ψ_s is zero. In this case there is no effective charge on gate electrode. Thus effective charge in surface layer Q_C is also zero. This condition is defined in (3.5).

$$V_{GB} = V_{FB}, \psi_S = 0, Q_C = 0 \tag{3.5}$$

In our further discussion it will be assumed that bulk is p-type semiconductor although using a n-type semiconductor for bulk also results in a similar device with a difference that charge carriers are of opposite polarity.

Now, if V_{GB} is smaller than V_{FB} then surface potential (ψ_S) becomes negative. The effective gate charge Q_G is also negative in this case so the surface has to be positively charged to balance it. Hence, holes in p-type bulk start accumulating near surface to make $Q_C > 0$ as shown in (3.6). This region of operation is known as accumulation region of MOS capacitor.

$$V_{GB} < V_{FB}, \psi_S < 0, Q_C > 0 \tag{3.6}$$

If V_{GS} is greater than V_{FB} then surface potential (ψ_S) becomes positive. The gate electrode is also positively charged. Thus surface layer has to become negatively charged to balance it. It does so by repelling the holes present in it and thus leaving it to be depleted of charge. Hence this region of MOS capacitor is known as depletion region. The concentration of electron at surface increases with increase in V_{GB}. When V_{GB} reaches to V_{L0}, surface potential becomes equal to fermi potential ϕ_F then concentration of electron at surface becomes equal to intrinsic concentration which is defined as boundary of depletion region. Thus, depletion region is defined as given in (3.7)

$$V_{L0} > V_{GB} > V_{FB}, \phi_F > \psi_S > 0, Q_C < 0 \tag{3.7}$$

When V_{GB} increases further causing an increase in ψ_S beyond ϕ_F then concentration of electron becomes higher than intrinsic concentration near surface. Thus surface start behaving like n-type semiconductor as if nature of semiconductor has changed or inverted from p-type to n-type. The layer formed due to negatively charged electrons near surface is known as inversion layer. This region of MOS capacitor operation is thus known as inversion region. The inversion region is divided in three sub regions named as weak inversion, moderate inversion and strong inversion depending upon surface potential. In the structure shown in Fig.3.3, increase in V_{GB} leads to band bending which results in flow of electrons from conduction band to region near surface. When V_{GB} reaches to V_{M0} then surface potential reaches to $\psi_S = 2\phi_F$ and concentration of electron in surface becomes equal to acceptor concentration in substrate. This is defined as upper boundary of the weak inversion region. This condition is also defined as threshold condition and hence gate voltage at threshold condition is defined as threshold voltage. It is given using (3.8)

$$V_t = V_{FB} + 2\phi_F + \psi_{ox}\ (\psi_S = 2\phi_F) \tag{3.8}$$

The weak inversion region is given by (3.9)

$$V_{M0} > V_{GB} > V_{L0},\quad 2\phi_F > \psi_S > \phi_F \tag{3.9}$$

In moderate inversion upper boundary of V_{GB} is V_{H0} which results in surface potential $\psi_S = 2\phi_F + \phi_Z$ Where ϕ_Z is several thermal voltage $\phi_t = \frac{kT}{q}$. Here, k is Boltzmann constant, T is absolute temperature in degree Kelvin and q is magnitude of the charge of an electron. The moderate inversion region in given by (3.10).

$$V_{H0} > V_{GB} > V_{M0},\quad 2\phi_F + \phi_Z > \psi_S > 2\phi_F \tag{3.10}$$

The strong inversion region is given by (3.11)

$$V_{GB} > V_{H0},\quad \psi_S > 2\phi_F + \phi_Z \tag{3.11}$$

We see that surface charge depends on surface potential and keeps on changing as surface potential changes. This can be effectively modelled using a surface capacitance C_C where this capacitance can be defined using (3.12)

$$C_C = \frac{\partial Q_C}{\partial \psi_S} \tag{3.12}$$

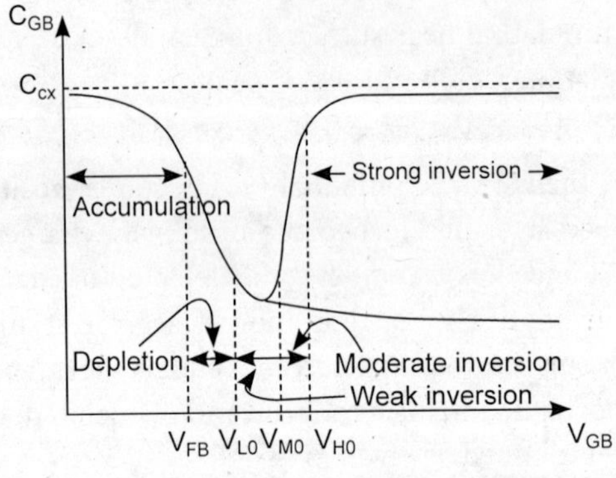

Fig. 3.4: Variation of MOS capacitance with external voltage

Here Q_C is surface charge and ψ_s is surface potential. The effective capacitance of MOS capacitor C_{GB} is essentially a series combination of oxide cap C_{ox} and surface capacitance C_C. Here C_{ox} is oxide capacitance of unit area. In accumulation region when V_{GB} is substantially smaller than V_{FB} then concentration of holes are significantly large near surface which effectively increases C_C substantially higher than C_{ox}. Thus C_{GB} approaches to C_{ox} as effective capacitance of capacitor connected in series is the smaller one. The surface charge reduces when V_{GB} approaches V_{FB} and even after when device enters in depletion region. Hence C_{GB} reduces. When device enters in strong inversion then again surface charge increases significantly resulting in increase of C_C and thereby C_{GB} taking it to C_{ox} again. The black curve shows the variation of C_{GB} with variation in V_{GB} at low frequency. The grey curve shows the variation of C_{GB} with variation in V_{GB} at high frequency. The formation of inversion layer is not possible at high frequency as it requires thermal generation and recombination which are rather slow process. Hence C_{GB} is rather small.

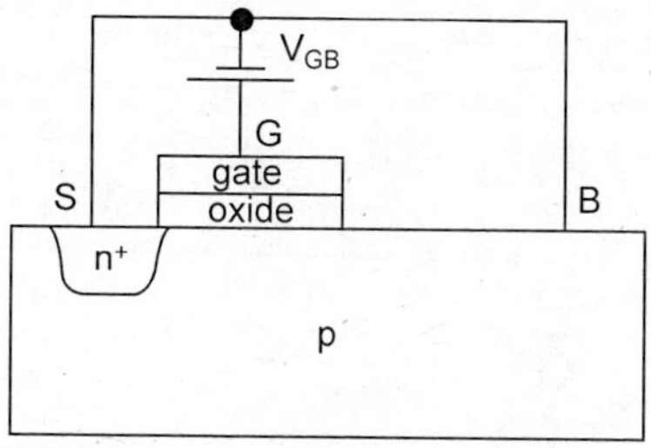

Fig. 3.5: MOS capacitor with source terminal

The Fig.3.5 shows a device similar to MOS capacitor discussed earlier with an added n$^+$ doped zone at left side of surface. This added n$^+$ zone act as a source of electrons needed to form inversion layer at surface hence it is also known as source. A contact terminal is provided to connect this n$^+$ doped zone to external world which is named as S because it belongs to source. The S and B terminals of this device are shorted so that p-n$^+$ junction remains deactivated. Having source close to surface allows electrons to quickly move to surface to create an inversion layer when V_{GB} is higher than V_t. Hence this device can have inversion layer even when V_{GB} is varied at a faster rate. Thus, this device does not follow different C-V characteristic if external voltage V_{GB} changes with different frequencies. It happens so because source provides electrons at a much faster rate than what is possible with thermal generation in earlier case. If VSB is nonzero and positive then source becomes positively charged and hence it start attracting electrons which reduces concentration of electrons in inversion layer. In fact increasing VSB beyond a value when VGB is fixed can make inversion layer disappear.

3.3.2 n-type MOS (NMOS)

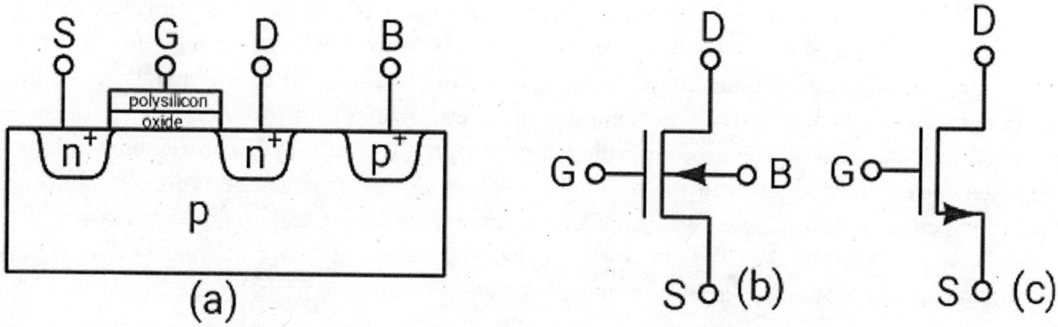

Fig. 3.6: (a) NMOS device (b) 4 terminal symbol of NMOS device (c) 3 terminal symbol of NMOS device

Adding another n$^+$ region to the right side of surface of device of Fig.3.5 creates a new device as shown Fig.3.6(a). The addition of this n+ doped zone allows possibility of a current flow between two n+ doped zones. The region between two n+ doped zone is defined as channel. The current flow through channel is possible when negatively charged inversion layer exist in surface. As the newly added n$^+$ region drains out electrons coming from source hence it is called as drain. This device is an n-type MOS transistor. It is termed as n-type MOS because flow of negatively charged electrons from source to drain results in a current flow through inversion layer. It is also called as NMOS transistor.

There are two possible structure for NMOS device. The device structure shown in Fig.3.6 (a) is an enhancement type NMOS transistor. This transistor is off when its gate to source voltage V_{GS} is below threshold voltage V_t. Thus V_{GS} of the device need to be higher than V_t to turn this device on.

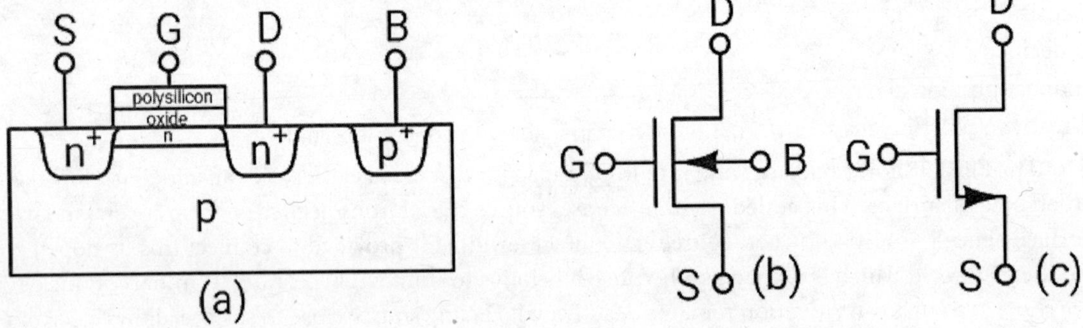

Fig. 3.7: (a) NMOS device (b) 4 terminal symbol of NMOS device (c) 3 terminal symbol of NMOS device

The Fig.3.7 shows another flavour of NMOS transistor where physically implanted n-type channel is present between source and drain. It allows the device to conduct even when its V_{GS} is zero. Actually a negative V_{GS} is needed to turn this device off. The negative V_{GS} creates a negative charge on gate electrode. This is mirrored by positive charged depleted ions in channel region which breaks the path between source and drain terminal of transistor. Thus depletion of channel region is responsible for turning the device off. Hence these devices are known as depletion type NMOS transistor. Such devices are used as a load resistor in power electronics applications.

NMOS transistor discussed here has four terminals named as source (S), Drain (D), gate (G) and bulk or body (B). The contact to bulk when using metal as contact wire is generally done through p+ region and not by directly connecting it to p region. This is done so because metal-semiconductor connection forms a Schottky diode if metal is connected directly to low doped semiconductor. The operation of MOS transistor is possible with access to only three terminals named source, drain and gate when body is shorted internally with source and thus this transistor can be a three terminal device. However when several NMOS transistors are realized on a same substrate then shorting source and body is not always possible and device essentially becomes a four terminal device. In most of the cases, body terminal is connected to an appropriate potential such that no current flows from body terminal to any other terminals.

The region of semiconductor between source and drain and below surface is referred as channel. The dimensions of this channel which are length(L) and width(W) of the channel are key design parameter of a MOS transistor. It can vary by order of magnitude based on design needs. The minimum channel length possible in a particular process is also used to define the process node such as 65nm CMOS process can have transistor with minimum channel length as 65nm.

The bulk terminal in MOS device is connected to a potential which effectively deactivates the p-n junction. In NMOS this is done by connecting bulk to minimum available supply. For a while now we will consider that source and bulk terminal are shorted and this shorted connection is maintained at minimum potential so that p-n junction is not forward biased.

In order to fabricate NMOS transistor, we begin by taking a p-type semiconductor which act as a substrate or body or bulk of this device. The two isolated buckets of heavily doped n-type semiconductor which forms n+ region are created in substrate using several different fabrication steps such as lithography, ion implantation and etching. These n+ regions form source(S) and drain(D) of the device. In a MOS transistor source and drain are interchangeable as they are identical and placed symmetrically. Its role is decided based on applied potential at these terminals. The contact to the substrate is provided by connecting a metal terminal to substrate through a heavily doped p+ region as connecting metal contact directly to substrate may result in formation of Schottky diode. It is known as Bulk or Body terminal (B). An oxide layer is deposited over the gap between source and drain region. The thickness of this oxide layer in advanced processes is around 1-2 nm. A low resistivity layer made up of polycrystalline silicon or metal is deposited on top of the oxide layer which act as a gate(G) of device. The interface of substrate and oxide layer is generally referred as surface.

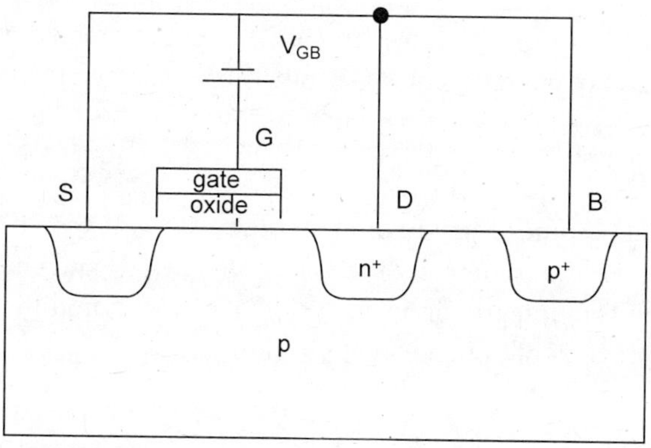

Fig. 3.8: MOS capacitor realized using an NMOS device

If S, D and B terminal of NMOS device are shorted together as shown in Fig.3.8 then it behaves like a two terminal MOS capacitor discussed earlier in this chapter. The voltage across the two terminal of this device is V_{GB} which is also equal to V_{GS} in this case. It forms an inversion layer when V_{GS} exceeds V_t. Thus, device is off when $V_{GS} < V_t$ and is said to be in cutoff region. Now if $V_{GS} \geq Vt$ then inversion layer forms and it can allow a current flow between its drain and source terminals. Now if a nonzero and positive V_{DS} is applied then an electric field is established in the channel between source and drain terminals of NMOS. This electric field allows electron coming from source to move to drain via inversion layer in channel and thus current flows from drain to source in this transistor.

In order to drive an expression of this current I_{DS}, we will first consider a scenario when V_{DS} is zero. In this case the channel charge is given using (3.13)

$$Q_{inv} = WC_{ox}(V_{GS} - V_t) \tag{3.13}$$

However when V_{DS} is nonzero then potential across channel varies from zero at source to V_{DS} at drain. Hence local potential difference between gate and channel is given using (3.14)

$$Q_{inv} = WC_{ox}(V_{GS} - V(x) - V_t) \tag{3.14}$$

Here $V(x)$ is local potential of the channel.

Hence I_{DS} is given using

$$I_{DS} = Q_{inv} \cdot v = WC_{ox}\,(V_{GS} - V(x) - Vt) \cdot \mu_n E \tag{3.15}$$
$$= W \cdot C_{ox} \cdot \left(V_{GS} - V(x) - V_t\right) \cdot \mu_n \frac{dV(x)}{dx}$$

Now on multiplying both side with dx and then integrating we get

$$\int_{x=0}^{x=L} I_{DS} dx = \int_{0}^{V_{DS}} W \cdot C_{ox} \cdot \left(V_{GS} - V(x) - V_t\right) \cdot \mu_n \cdot dV(x) \tag{3.16}$$

As I_{DS} is constant along the channel so (3.16) is simplified as (3.17)

$$I_{DS} = \mu_n C_{ox} \frac{W}{L}\left((V_{GS} - V_t)V_{DS} - \frac{V_{DS}^2}{2}\right) \tag{3.17}$$

For small value of V_{DS} expression shown in (3.17) is approximated as (3.18) which is a linear expression. Hence this region of operation of MOS transistor is also referred as linear region.

$$I_{DS} = \mu_n C_{ox} \frac{W}{L}(V_{GS} - V_t)V_{DS} \tag{3.18}$$

Now, If drain potential becomes equal to or higher then $V_{GS} - V_t$ then inversion layer ceases to end at drain side of the channel as local potential difference between gate and channel at drain end becomes smaller than V_t. Hence, limits of integration in right side of (3.16) is modified to $V_{DS} = 0$ to $V_{DS} = V_{GS} - V_t$. Using it we get the expression of drain current given in (3.19)

$$I_{DS} = \frac{1}{2}\mu_n C_{ox} \frac{W}{L}(V_{GS} - V_t)^2 \tag{3.19}$$

Now I_{DS} no longer depends on V_{DS} so it does not change when V_{DS} is increased beyond $V_{GS} - V_t$ as if it is saturated. Hence this region of operation of MOS transistor is known as saturation region.

However if V_{GS} becomes smaller than V_t then inversion layer ceases to end throughout the channel and hence I_{DS} becomes zero in this case. Table.3.1 summarizes this discussion and provides details about operating points for different operating region of NMOS transistor.

3.3.3 Operating Regions of NMOS Device

S.No.	Operating region	Condition on V_{GS}	Condition on V_{DS}	I_{DS}
1	Cutoff	$V_{GS} < V_t$		0
2	Saturation	$V_{GS} \geq V^t$	$V_{DS} \geq V_{GS} - V_t$	$\frac{1}{2}\mu_n C_{ox} \frac{W}{L}(V_{GS} - V_t)^2$
3	Linear	$V_{GS} \geq V_t$	$V_{GS} - V_t > V_{DS} > 0$	$\mu_n C_{ox} \frac{W}{L}\left((V_{GS} - V_t)V_{DS} - \frac{V_{DS}^2}{2}\right)$

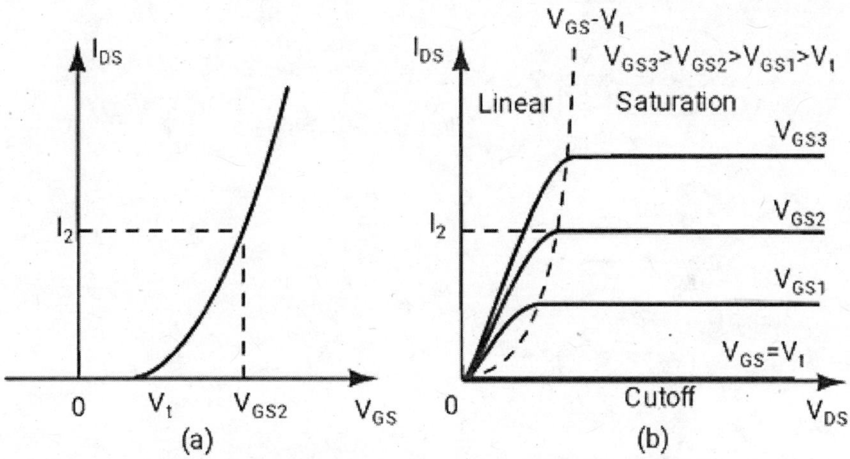

Fig. 3.9: (a) Input characteristic of NMOS device (b) Output characteristic of NMOS device

Fig. 3.9 (a) shows the input characteristic of an NMOS device. The device is off when its gate to source voltage V_{GS} is smaller than its threshold voltage V_t. This region of operation of NMOS device is known as cutoff region. The device act like an open switch between its drain and source terminal when operated in this region. When V_{GS} is larger than V_t then device turns on in a way that current starts flowing between its drain and source terminal. If its drain to source voltage V_{DS} is less than $V_{GS} - V_t$ then its drain current is governed by (3.17) where for small values of V_{DS}, $I_{DS} - V_{DS}$ characteristic which is also known as output characteristic and is shown in Fig. 3.9 (b) is linear function of V_{DS}. Therefore, this region of operation is known as linear region. The device behaviour in this region is like a resistor where current flowing between its drain and source terminal is linear function of voltage across drain and source terminal. Hence this device is used to realize resistor where its resistance is given using (3.20).

$$R_{on} = \frac{1}{\mu_n C_{ox} \frac{W}{L}(V_{GS} - V_t)} \tag{3.20}$$

This resistance is a function of V_{GS} and hence it is a voltage controlled resistor. If V_{GS} is high (This high value is generally V_{DD} which is highest possible supply in most of the circuits) then resistance is much smaller and device acts like a closed switch.

For values of V_{DS} higher than $V_{GS} - V_t$, the output characteristic becomes constant for a specific V_{GS} and does not depend on V_{DS}. Hence this region is known as saturation region of MOS transistor operation. A MOS transistor can act as a current source or more specifically as a voltage controlled current source when operated in this region.

3.3.4 Application of NMOS as a Switch

Thus to operate NMOS device as a switch one can operate it either in cutoff region where it acts like an open switch or in linear region where it acts like a closed switch. The switch gets opened when $V_{GS} = 0$ as shown in Fig. 3.10 (a) and switch gets closed when $V_{GS} = V_{DD}$ as shown in Fig. 3.10 (b). This operation is very useful in the context of digital circuit design where binary operations are performed using this switch.

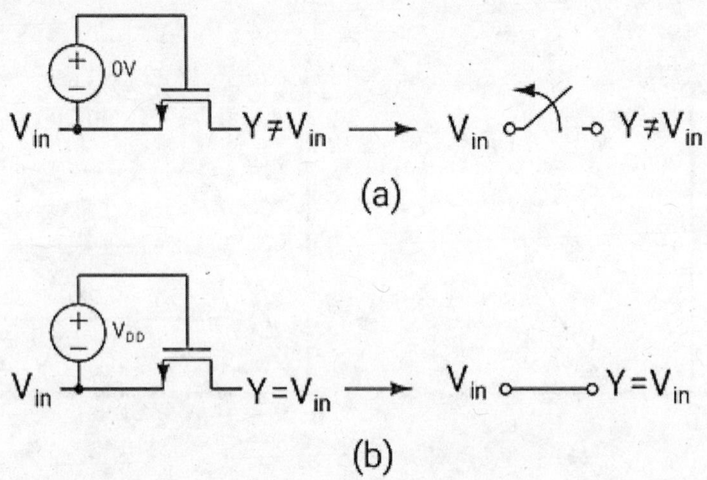

Fig. 3.10: (a) NMOS switch in off state when its V_{GS} is low
(b) NMOS switch in its on state when its V_{GS} is high

NMOS device when operated in its saturation region of operation acts like a current source as it provides a constant current even when output voltage across it VDS is varied. Actually it acts like a voltage controlled current source where its drain current I_{DS} becomes a function of its gate to source voltage V_{GS}. The device is used as an amplifier when operated in this region as it can provide a very high small signal output impedance when operated in this region.

3.3.5 Channel length modulation

So far it is assumed that drain current of an NMOS transistor is only a function of its gate to source voltage V_{GS} when operated in saturation region. However effective length of channel changes with V_{DS} after pinchoff has occurred which is the case with saturation region. Hence actual current of an NMOS transistor in saturation region is better described using expression shown in (3.21).

$$I_{DS} = \frac{1}{2}\mu_n C_{ox}\frac{W}{L}(V_{GS} - V_t)^2(1 + \lambda V_{DS}) \tag{3.21}$$

Here λ is channel length modulation parameter. It is defined as $\lambda = \frac{\Delta L}{L}$ where ΔL is length of pinched off channel and L is total channel length of device. The equation (3.21) shows that drain current of a transistor is a function of V_{GS} and V_{DS}. However, it is a strong function of V_{GS} and a weak function of V_{DS}.

3.3.6 Small Signal Model of NMOS Transistor

The small signal model of NMOS transistor is developed by following steps similar to (1.6). The gate current in NMOS is 0 and drain current of NMOS device I_{DS} is defined by (3.21). Hence small signal equations relating small signal current of NMOS transistor with small signal voltages of NMOS transistor are given in (3.22) and (3.23).

$$\Delta i_g = 0 \tag{3.22}$$

$$\Delta i_d = \frac{\partial I_{DS}}{\partial V_{GS}}\Big|_{(I_{DS},V_{GS},V_{DS})} \Delta v_{gs} + \frac{\partial I_{DS}}{\partial V_{DS}}\Big|_{(I_{DS},V_{GS},V_{DS})} \Delta v_{ds} \tag{3.23}$$

$$= g_m \, \Delta v_{gs} + g_o \, \Delta v_{ds}$$

Here Δi_g, Δi_d, Δv_{ds} Δv_{gs} are small signal or incremental changes in gate current I_G, drain current I_{DS}, gate to source voltage V_{GS} and drain to source voltage V_{DS} of NMOS transistor respectively. The equation (3.23) suggest that Δi_d is a function of two small signal parameters. Here parameter g_m is the variation of drain current I_{DS} with gate to source voltage V_{GS} evaluated at operating points. It is essentially a voltage controlled current source connected between drain and source node of NMOS transistor.

The expression of g_m for NMOS is calculated using (3.21) and (3.23) and given in (3.24). One can generally assume $\lambda = 0$ while calculating g_m without much loss of error as it simplifies the calculations.

$$g_m = \frac{\partial I_{DS}}{\partial V_{GS}}\Big|_{(I_{DS},V_{GS},V_{DS})} = \frac{2I_D}{(V_{GS} - V_t)} = \mu_n C_{ox} \frac{W}{L}(V_{GS} - V_t) = \sqrt{2I_D \mu_n C_{ox} \frac{W}{L}} \tag{3.24}$$

Any of the three expressions can be used to calculate g_m depending upon the known parameters or operating points.

Similarly, g_o is the variation of drain current I_{DS} with drain to source voltage V_{DS} evaluated at operating points. It is essentially a resistor between drain and source terminals of transistor. This is so because current between the terminals is proportional to voltage across it which is the case with resistor. The Fig.3.11 shows the small signal model of an NMOS transistor.

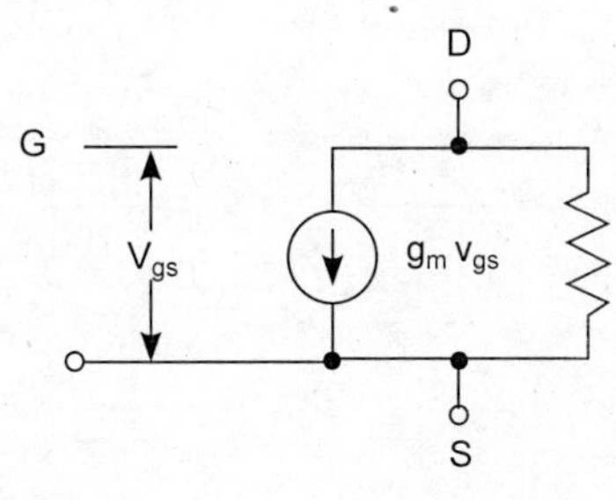

Fig. 3.11: Small signal model of NMOS transistor

The expression of g_o for NMOS is calculated using (3.21) and (3.23) and given in (3.25).

$$g_o = \frac{\partial I_{DS}}{\partial V_{DS}}\Big|_{(I_{DS},V_{GS},V_{DS})} = \lambda I_{DS} \tag{3.25}$$

Example:3.1:

Calculate the operating points of circuit shown in Fig.3.12. Find its small signal parameters? Consider $V_t = 1\,V$ and $\mu_n \, C_{ox} = 200\mu A/V^2$.

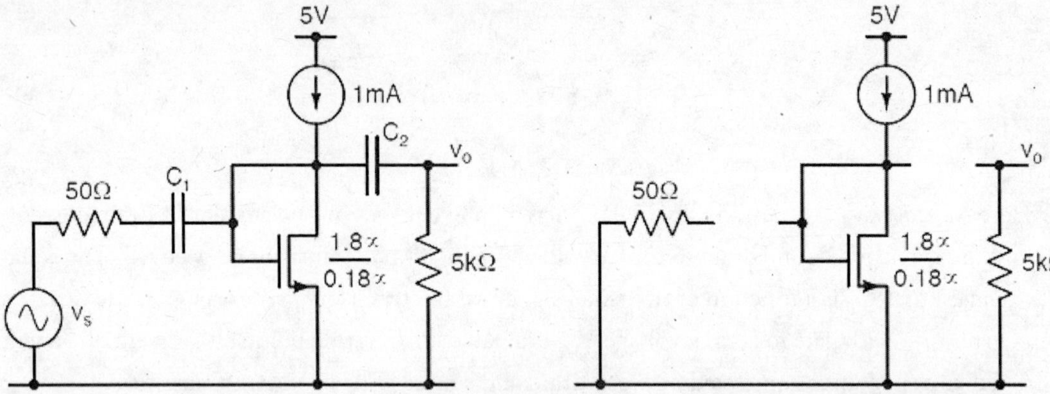

Fig. 3.12: (a) Circuit diagram for example: ;3.1 (b) DC equivalent circuit of Fig.3.12 (a)

Solution:

In order to find the DC operating point of the circuit shown in Fig.3.12(a) we first draw its DC equivalent circuit. The DC equivalent circuit is obtained by performing following changes in original circuit: (a) Capacitors are made open in DC equivalent circuit as it offer an infinite impedance at DC. Similarly if there is any inductor in the circuit then it is shorted as inductor has zero impedance at DC. (b) All AC independent sources present in circuits are nulled which means all independent AC voltage sources are replaced with short circuit and all independent AC current sources present in circuit are replaced with open circuit. The Fig.3.12(b) shows the DC equivalent circuit of circuit shown in Fig.3.12(a).

Now DC equivalent circuit is solved. As per KCL current through transistor I_{DS} is same as current in current source which is equal to 1mA. Further, as gate and drain are at same potential and current through transistor is nonzero therefore transistor must be in saturation region. Hence V_{GS} is calculated using (3.26).

$$V_{GS} = \sqrt{\frac{2I_D}{\mu_n C_{ox}\frac{W}{L}}} + V_t = \sqrt{\frac{2m}{200\mu \times 10}} + 1 = 1 + 1 = 2V \tag{3.26}$$

As gate and source of transistor are shorted hence V_{DS} will also be same as V_{GS}. Now g_m is evaluated using (3.27).

$$g_m = \frac{2I_D}{V_{GS} - V_t} = \frac{2mA}{(2-1)V} = 2mS \tag{3.27}$$

g_o is evaluated using (3.28)

$$g_o = \lambda I_D = 0.1V^{-1} \times 1\text{ m A} = 0.1\text{ mS} \tag{3.28}$$

Inverse of g_o is termed as r_o which is small signal output impedance of NMOS. Hence it is given using (3.29)

$$r_o = \frac{1}{\lambda I_D} = \frac{1}{0.1mS} = 10k\Omega \tag{3.29}$$

3.4 AMPLIFIER CONFIGURATIONS

Like BJT, MOS device characteristic is also suitable for designing amplifiers. As amplifier has two ports which require two terminals each and hence one terminal is shared between input and output port while designing an amplifier using a three terminal device like transistor. Hence there are three possible amplifier configuration while designing it using an NMOS transistor. These amplifier configurations are discussed below.

3.4.1 Common Source Amplifier

If the source terminal is shared between input and output port then designed amplifier is known as common source (CS) amplifier.

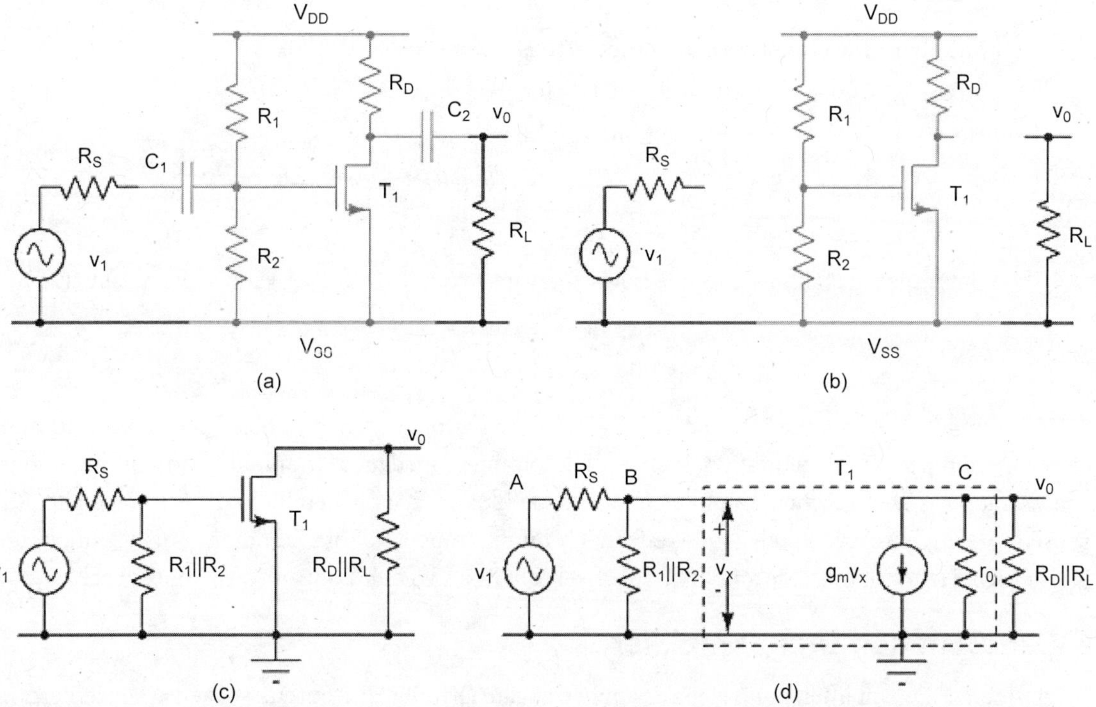

Fig. 3.13: (a) CS amplifier circuit (b) DC equivalent circuit of Fig.3.13(a) (c) Small signal equivalent circuit of Fig.3.13(a) (d) Small signal equivalent circuit of CS amplifier after substituting small signal model of NMOS transistor

The Fig.3.13(a) shows a common source amplifier designed using NMOS transistor. The part of the circuit shown in blue is essentially the biasing circuit which ensures that NMOS (T_1) is biased in saturation region of NMOS transistor operation. The values of components are chosen based on expected small signal parameters. The capacitors shown in red ensures that DC bias points of T_1 are not affected by the small signal source and load. As capacitors are open at DC, therefore, source and load are effectively disconnected from the transistor circuit. These capacitors are termed as coupling capacitor and its value is chosen is such a way that at signal frequencies these capacitors effectively get shorted. Fig. 3.13 (b) shows the DC equivalent circuit of common source amplifier which is used to evaluate the operating points of devices. Fig.3.13(c) shows the small signal equivalent of CS amplifier circuit shown in Fig. 3.13 (a) where coupling capacitors $C_{1,2}$ are shorted as its impedance at signal frequencies are low. Now transistor can also be replaced with its small signal equivalent circuit which brings us to Fig. 3.13 (d). The circuit shown in Fig.3.13 (d) is analyzed to find the small signal performance of this circuit.

The small signal gain of common source amplifier is obtained by solving its equivalent small signal circuit shown in Fig.3.13(d).

Node potential at node B which is v_x is obtained by solving the resistive divider at input side. It is given by (3.30).

$$v_x = \frac{(R_1 \| R_2)}{R_S + (R_1 \| R_2)} v_i \tag{3.30}$$

Now on solving the circuit at output node, v_o is given using (3.31)

$$v_o = -g_m v_x (R_D \| R_L \| r_o) \tag{3.31}$$

Using (3.30) and (3.31) we can obtain the amplifier gain given by (3.32)

$$A_v = \frac{v_o}{v_i} = -g_m \frac{(R_1 \| R_2)}{R_S + (R_1 \| R_2)} (R_D \| R_L \| r_o) \tag{3.32}$$

If output impedance of input source is low then approximate amplifier gain is given using (3.33)

$$A_v = \frac{v_o}{v_i} = -g_m (R_D \| R_L \| r_o) \tag{3.33}$$

The impedance looking into gate terminal of transistor very high. Hence the input impedance of a common source configuration is high. The output impedance of amplifier in common source configuration is $R_D \| r_o$ when amplifier is not loaded. Here, r_o is generally high resistance and R_D is chosen to be high to have a high gain so effectively output impedance of a common source configuration is also high. Hence in this case amplifier effectively act like a voltage controlled current source.

3.4.2 Common Drain Amplifier

In common drain amplifier, input signal is applied at gate terminal and output is taken at source terminal of transistor while drain terminal is biased using constant DC potential. The amplifier circuit designed using this configuration is shown in Fig.3.14(a). Its DC equivalent circuit is shown in Fig.3.14(b). Here current source I_o is used to bias the transistor T_1. It fixes drain current to I_o ensuring that biasing is robust. The small signal equivalent circuit shown in Fig.3.14 (c) is obtained by shorting the coupling cap and nullifying the DC sources present in circuit. Finally on substituting the small signal model of transistor we get the complete small signal equivalent circuit which is shown in Fig.3.14(d).

Now to calculate the small signal gain of this circuit one can first find potential at node B using voltage divider

$$v_B = \frac{R_1 \| R_2}{R_S + R_1 \| R_2} v_i \approx v_i \tag{3.34}$$

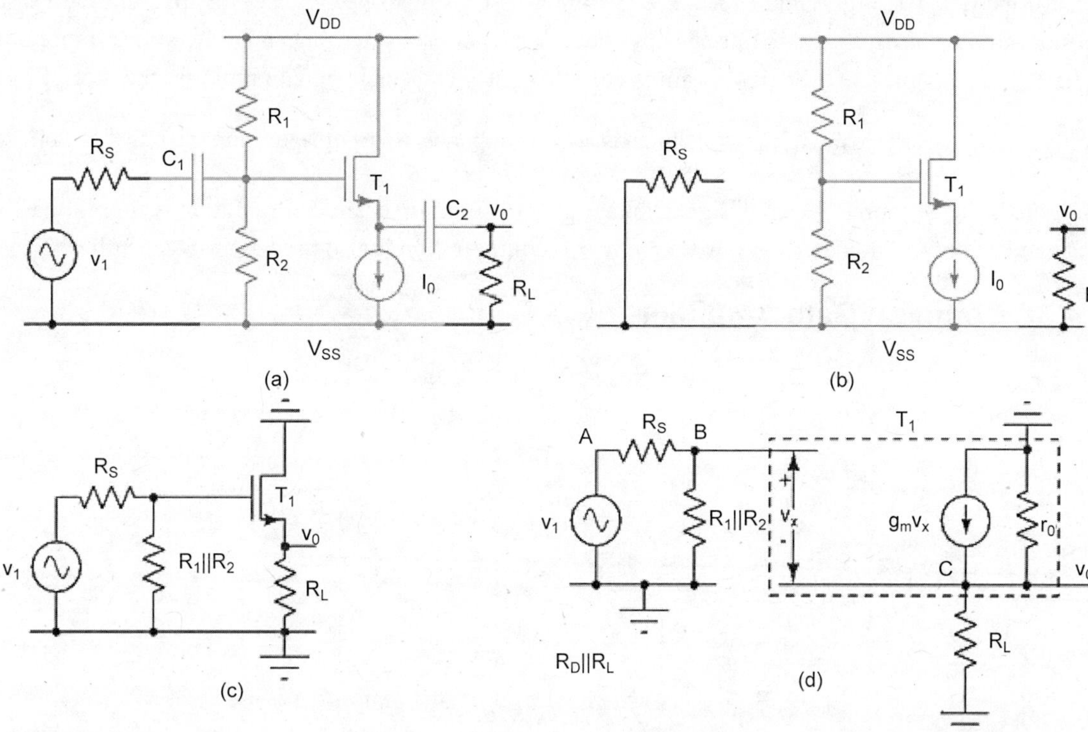

Fig. 3.14: (a) CD amplifier circuit (b) DC equivalent circuit of Fig.3.14(a) (c) Small signal equivalent circuit of Fig.3.14(a) (d) Small signal equivalent circuit of CD amplifier after substituting small signal model of NMOS transistor

Now writing KCL at output node of the circuit shown in Fig.3.14(d)

$$g_m\left(v_B - v_o\right) = \frac{v_o}{R_L \,||\, r_o} \tag{3.35}$$

On simplifying (3.35) we obtain (3.36)

$$v_o = \frac{v_B(R_L \,||\, r_o \,||\, \frac{1}{g_m})}{(\frac{1}{g_m})} = \frac{g_m(R_L \,||\, r_o)v_B}{1 + g_m(R_L \,||\, r_o)} \tag{3.36}$$

Using (3.35) and (3.36) we obtain (3.37)

$$\frac{v_o}{v_i} = \frac{g_m(R_L \,||\, r_o)v_B}{1 + g_m(R_L \,||\, r_o)} \approx 1 \tag{3.37}$$

The small signal input impedance in common drain configuration is the current provided by test source when it is connected at its input terminal which is gate terminal in this case. It is given by (3.38) for the amplifier shown in Fig.3.14(a)

$$Z_{in} = R_S + (R_1||R_2) \tag{3.38}$$

It is generally a large value as $R_{1,2}$ are chosen large resistor to keep power dissipation low. These resistors are flexible in a way that its absolute value can be chosen anything and only ratio is important to fix the bias point. Its small signal gain is close to unity. The small signal output impedance of this configuration is approximately $\frac{1}{g_m}$ which is low and hence this configuration acts like a voltage controlled voltage source. As small signal voltage gain of common drain configuration is approximately unity and its output impedance is low therefore this configuration is also used as a voltage buffer.

3.4.3 Common Gate Amplifier

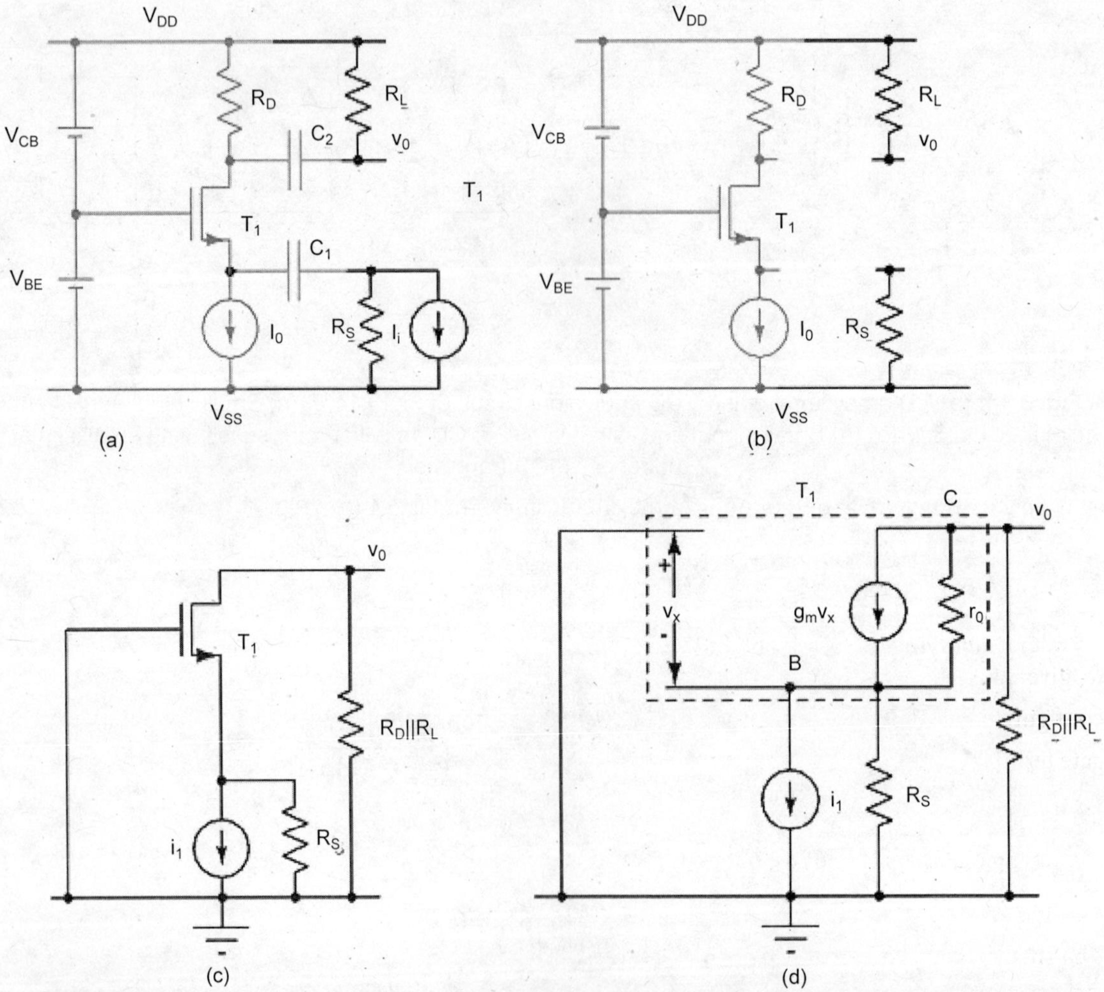

Fig. 3.15: (a) CG amplifier circuit (b) DC equivalent circuit of Fig.3.15(a) (c) Small signal equivalent circuit of Fig.3.15(a) (d) Small signal equivalent circuit of CG amplifier after substituting small signal model of NMOS transistor

In common gate amplifier, input signal is applied at source terminal and output is taken from drain terminal of transistor while gate terminal is biased using constant DC potential. The amplifier circuit designed using this configuration is shown in Fig.3.15(a). Its DC equivalent circuit is shown in Fig.3.15(b). Here current source I_o is used to bias the transistor T_1. It fixes drain current to I_o and thus this biasing is robust. The small signal equivalent circuit shown in Fig.3.15(c) is obtained by shorting the coupling cap and nullifying the DC sources present in circuit. Finally on substituting the small signal model of transistor we get the complete small signal equivalent circuit which is shown in Fig.3.15(d).

Now to calculate the small signal gain of this circuit one can write KCL at the node B of the circuit shown in Fig.3.15(d) which is given in (3.39)

$$i_i - \frac{v_x}{R_S} - g_m v_x - \frac{v_x + v_o}{r_o} = 0 \tag{3.39}$$

Now writing KCL at output node of the circuit shown in Fig.3.15(d)

$$g_m v_x + \frac{v_o + v_x}{r_o} + \frac{v_o}{R_D \,||\, R_L} = 0 \tag{3.40}$$

On simplifying (3.40) using we obtain (3.41)

$$v_x = \frac{-v_o r_o}{(R_D \,||\, R_L \,||\, r_o)(1 + g_m r_o)} \tag{3.41}$$

Using (3.40) and (3.41) we obtain (3.42)

$$\frac{v_o}{i_i} = \frac{-(R_D \,||\, R_L \,||\, r_o)(R_S \,||\, \frac{1}{g_m} \,||\, r_o)(1 + g_m r_o)}{r_o} \approx -(R_D \,||\, R_L \,||\, r_o) \tag{3.42}$$

The small signal current gain of CG amplifier is essentially unity which flows into output impedance to give small signal gain expressed in (3.42). The small signal input impedance in common drain configuration is the current provided by test source when it is connected at its input terminal which is gate terminal in this case. It is given by (3.43) for the amplifier shown in Fig.3.15(a)

$$Z_{in} = R_S + \frac{r_o + (R_D \,||\, R_L)}{g_m r_o + 1} \approx R_S + \frac{1}{g_m} \tag{3.43}$$

It is generally a small value as source series resistance R_S is generally small and g_mis large quantity. The small signal output impedance of this amplifier is given in (3.44) which is a large quantity. Hence, CG amplifier acts like a current buffer.

$$Z_{out} = (R_S(g_m r_o + 1) + r_o) \,||\, R_D \tag{3.44}$$

3.5 P-TYPE MOS (PMOS)

Fig.3.16 (a) shows a cross sectional diagram of p-channel MOSFET or PMOS. It is an enhancement type PMOS device. It has four terminals named as source (S), Drain (D), Gate (G) and Bulk (B). The current in PMOS flows from source to drain terminal. The Fig.3.16(b) shows a 4-terminal symbol of enhancement type PMOS device and Fig.3.16(c) shows a 3-terminal symbol of an enhancement type PMOS device. The broken line symbol represent that channel is not present when no external voltage is applied to the device. The symbol of depletion type PMOS device does not have broken lines in its symbol which represent that its channel is present when no external voltage is applied to it. However in general practice we draw a continuous line symbol even for enhancement type PMOS for simplicity as it is easy to draw. Further, in most of the cases circuit has either enhancement type or depletion type MOS transistor so same symbol is used without creating a confusion.

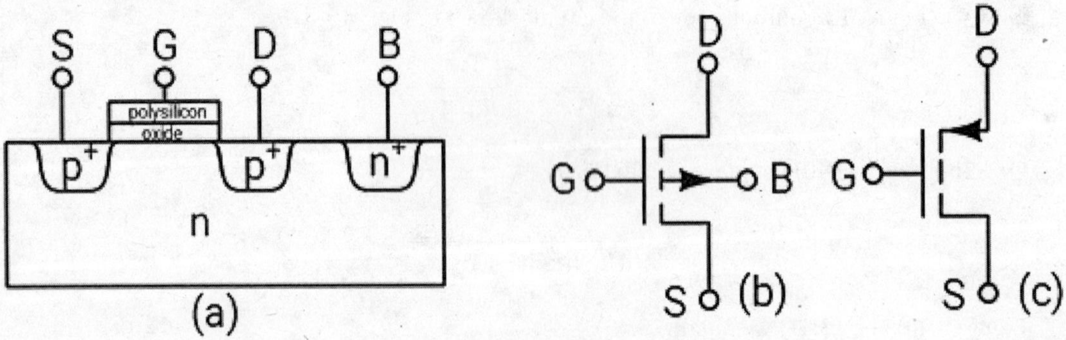

Fig. 3.16: (a) PMOS device (b) Four terminal symbol of PMOS (c) Three terminal symbol of PMOS

PMOS is a dual of NMOS device and thus holes are responsible for current flow in PMOS. The Tab.3.2 provides the details of condition on operating points for operating in different region of operation of PMOS transistor. These conditions are similar to the condition shown in Tab.3.2 with few differences such as mobility of charge carrier is different as here holes are involved instead of electrons which have reduced mobility. Hence PMOS device of same size is slower than NMOS device.

Tab.3.2: Details on operating conditions for operating in different region of operation of PMOS transistor

S.No.	Operating region	Condition on V_{GS}	Condition on V_{DS}	I_{DS}
1	Cutoff	$V_{SG} < \|V_t\|$		0
2	Saturation	$V_{SG} \geq \|V_t\|$	$V_{SD} \geq V_{SG} - \|V_t\|$	$\frac{1}{2}\mu_p C_{ox} \frac{W}{L}(V_{SG} - \|V_t\|)^2$
3	Linear	$V_{SG} \geq \|V_t\|$	$V_{SG} - \|V_t > V_{SD} > 0$	$\mu_p C_{ox} \frac{W}{L}\left((V_{SG} - \|V_t\|)V_{SD} - \frac{V_{SD}^2}{2}\right)$

The small signal model of PMOS is derived using the steps similar to the one followed while deriving small signal model of NMOS transistor. It is exactly same as small signal model of NMOS device.

3.6 UNIJUNCTION TRANSISTOR (UJT)

As the name implies unijunction transistor (UJT) is a three terminal device with a single p-n junction. The p-region is doped relatively higher than n region as this region is used to inject or emit charge in n-region. Hence this region is also called as emitter. The n-region is also termed as base region of this device. Fig.3.17(a) shows the cross section of a unijunction transistor. Fig.3.17(b) uses this UJT in a circuit and Fig.3.17(c) shows its effective model to analyze its operation.

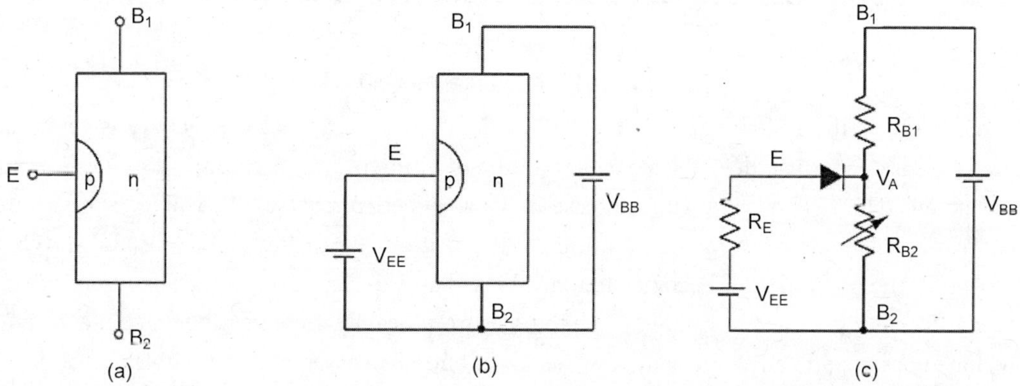

Fig. 3.17: (a) UJT device (b) Circuit designed using UJT (c) Circuit of Fig.3.17(b) with equivalent model of UJT

In UJT, n-type region is lightly doped and act as a resistive path between its two terminal B_1 and B_2. Here RB_1 is effective resistance of base region from top end (B_1) to junction end and RB_2 is effective resistance of base region from junction end to bottom end (B_2). Hence potential at junction is defined using (3.45)

$$V_A = \frac{R_{B2}}{R_{B1} + R_{B2}} V_{BB} \tag{3.45}$$

Now, voltage needed to forward bias p-n junction in this device is approximately $V_A + V_D$ where V_D is drop across diode when it is forward biased. Thus as soon as V_E reaches to $V_E = V_A + V_D$, junction becomes forward biased. This results in injection of holes into n-type base region from emitter which finally moves toward lower end of base region B_2. These holes coming from emitter has longer life time as base is lightly doped and thus results in increase in amount of charge in region between junction and lower end of base region. This effectively reduces resistance R_{B2}. The reduction in R_{B2} effectively reduces V_A which increases magnitude of forward bias. Thus this process is regenerative and results in sharp drop in V_A. However at certain point the concentration of holes saturates and resistance R_{B2} does not decrease any further. Now any further increase in current results in increase in V_A.

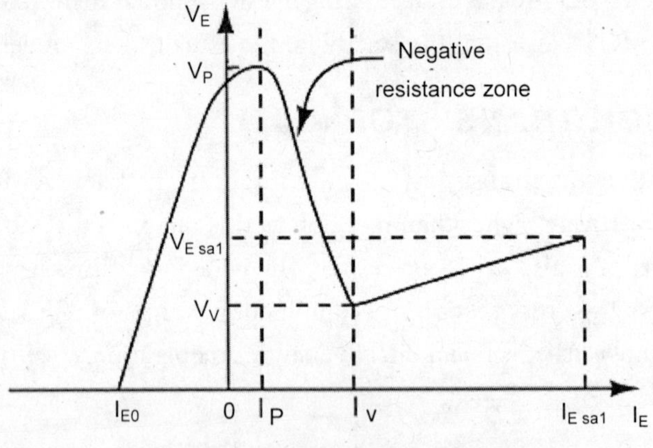

Fig. 3.18: I-V characteristic of UJT

Fig.3.18 shows the I-V characteristic of UJT. Initially when external voltage V_{EE} is zero, emitter voltage V_E is also zero. The device is considered off and emitter current I_E is mostly due to minority charge carriers. The I_E decreases with increase in V_E as effective reverse bias voltage across junction reduces. However the device remains off untill emitter voltage V_E reaches to $V_P = V_A + V_D$. When V_E reaches to VP the p-n junction becomes forward biased and device turns on causing emitter current to increase. The emitter current is largely due to holes from heavily doped emitter region entering into lightly doped base region and especially towards lower end of base region. These holes have longer life time in lightly doped base region and thus decreases value of R_{B1}. The emitter voltage V_E also decreases with decrease in R_{B1}. Hence I-V characteristic of device shows a negative slope. Thus UJT shows an incremental negative resistance when operated in this region however, overall resistance is still positive. This trend continues till emitter voltage reaches to VV where base region is saturated with holes injected by emitter and R_{B1} does not decrease any further. Now increase in V_{EE} increases I_E but also increases V_E. However, the device resistance is rather low. Now in order to turn off device, as one reduces V_{EE}, emitter current I_E also reduces. When I_E becomes smaller than IV, I-V characteristic of UJT switches back to its off region.

UNIT SUMMARY

This chapter has provided an introduction to basic operation of field effect transistor (FET) as well as its semiconductor physics. It has also covered different types of FET and provided a comprehensive analysis of its operational differences. The basics of MOS transistor which is a type of FET has been covered in detail. It has also provided a detailed analysis of MOS capacitor. It has also provided steps to analyze small signal behaviour of circuits designed using MOS transistor. The chapter then provides design details about different configuration of amplifier realized using MOS transistor. Description of unijunction transistor is provided to complete the discussion on FET.

EXERCISES

Multiple Choice Questions

1. Which of the following device is a unipolar device?
   (a) BJT
   (b) MOSFET
   (c) TRIAC
   (d) IGBT
2. Which of the following device is not a passive device?
   (a) Resistor
   (b) Diode
   (c) FET
   (d) LED
3. Which of the following statement is correct?
   (a) FET is a current controlled device and has high input impedance
   (b) FET is a voltage controlled device and has high input impedance
   (c) FET is a current controlled device and has low input impedance
   (d) FET is a voltage controlled device and has low input impedance
4. Which of the following is a voltage controlled device
   (a) Silicon controlled rectifier
   (b) Bipolar junction transistor
   (c) Insulated gate Bipolar transistor
   (d) Field effect transistor
5. Which of following amplifier configuration has least output impedance
   (a) Common source amplifier
   (b) Common drain amplifier
   (c) Common gate amplifier
   (d) Source degenerated common source amplifier
6. Which of the following region of operation of MOS devices are used to operate MOS as a switch?
   (a) Cutoff and linear
   (b) Cutoff and saturation
   (c) Linear and saturation
   (d) All three regions
7. Behaviour of MOSFET is approximated as which of the following device when operated in linear region
   (a) Voltage controlled resistor
   (b) Voltage controlled current source
   (c) Voltage controlled inductor
   (d) Current controlled current source
8. Which of the following amplifier configuration is used as a voltage buffer
   (a) Common source amplifier
   (b) Common drain amplifier
   (c) Common gate amplifier
   (d) Source degenerated common source amplifier
9. Which of the following amplifier configuration is used as a current buffer
   (a) Common source amplifier
   (b) Common drain amplifier
   (c) Common gate amplifier
   (d) Source degenerated common source amplifier

10. Body terminal of an NMOS device is always connected to

(a) Smallest available potential
(b) Highest available potential
(c) Can be connected to any potential
(d) Zero potential

11. The resistance of a MOSFET can be reduced by performing which of the following modification.

(a) Increasing its length
(b) Reducing its width
(c) Increasing its threshold voltage
(d) Increasing its width

ANSWERS															
1.	(b)	2.	(c)	3.	(b)	4.	(d)	5.	(b)	6.	(a)	7.	(b)	8.	(b)
9.	(c)	10.	(a)	11.	(d)										

Short Answer Type Questions

1. What is the difference between BJT and FET?
2. What is the difference between JFET and MOSFET?
3. What are different region of operation in a MOS transistor? What are the operating condition for these region of operations?
4. Which region of operation should be selected for a MOSFET to operate it as a current source?
5. Which region of operation should be selected for a MOSFET to operate it as a controlled resistor?
6. What are the different amplifier configuration designed using MOS transistor?
7. What happens to output impedance of MOS transistor when a smaller process node is used?
8. Why gain of MOS transistor becomes an issue in an advanced process nodes?
9. Which of the MOS amplifier configuration has least output impedance?
10. Which of the MOS amplifier configuration has higher output impedance?
11. What is the effective width of MOS device when it is connected in parallel if width of individual device is W?

Long Answer Type Questions

1. Define the operation of MOS capacitor. Plot the curve of capacitance as a function of .
2. What is inversion layer? How it gets form in a MOS capacitor? Why it is called as inversion layer?
3. Which of the operating region of MOS device is used for designing amplifier? What are the operating condition for MOS device to operate in this region?
4. Compare the different configuration of MOS amplifier.
5. Derive the expression of small signal gain of a common source amplifier which is loaded with an Ideal current source of value I at its drain node.
6. A MOS device is carrying a constant current and operates in saturation region of operation. Plot its intrinsic gain as a function of W/L.
7. Derive the expression of output impedance of a MOS transistor using (3.21).

Numerical Problems

1. Find the current in an NMOS transistor if it is operating in saturation region and it is given then $\mu_n C_{ox} = 100\mu A/V^2$, W = 1.8μm, L = 0.18μm, $V_t = 1V$ and $V_{GS} = 2V$. λ can be assumed zero here. Also find the transconductance of the device.
2. For the device of Q.N3.1 if $\lambda = 0.1V^{-1}$ then find the intrinsic gain of the NMOS transistor.
3. Find the operating points of the circuit shown in Fig.3.19. Also find its small signal gain. it is given then $\mu_n C_{ox} = 100\mu A/V^2$, W = 2μm, L = 1μm, $V_t = 1V$ and $V_{GS} = 2V$. λ can be assumed zero here. Assume C is very large capacitance.

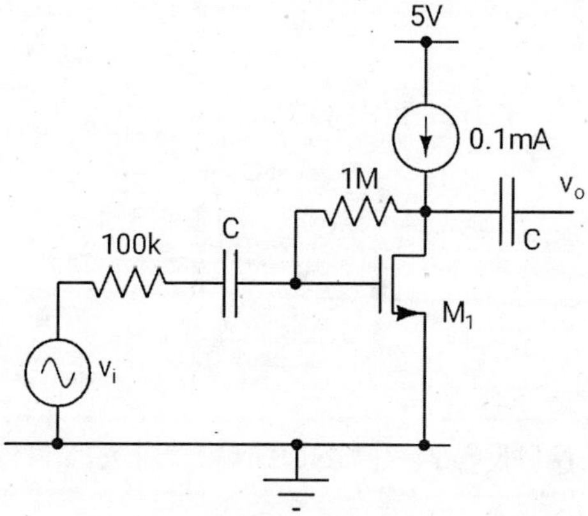

Fig. 3.19: Circuit for Problem Q.N3.3

4. Find the operating points of the circuit shown in Fig.3.20. Also find its small signal gain. it is given then $\mu_n C_{ox} = 100\mu A/V^2$, W = 2μm, L = 1μm, $V_t = 1V$ and $V_{GS} = 2V$. λ can be assumed zero here. Assume C is very large capacitance.

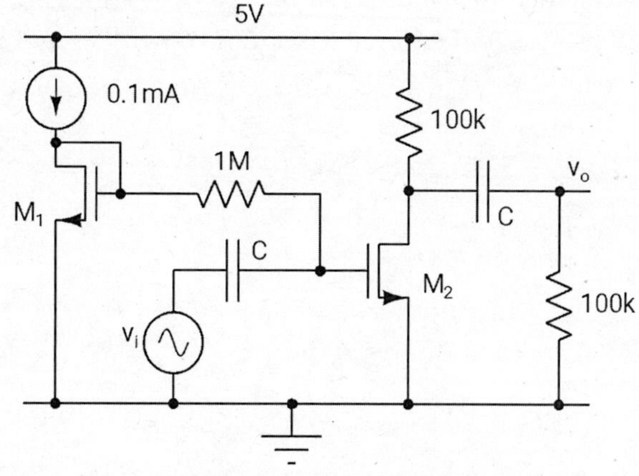

Fig. 3.20: Circuit for Problem Q.N3.4

KNOW MORE

The fabrication of MOS transistor is currently done using CMOS process which allows the fabrication of NMOS and PMOS transistor on a single substrate of silicon material along with several other components such as resistors, capacitors. This way whole circuits are designed on a single chip. The CMOS process has significantly benefitted from process scaling. The process scaling is the trend in semiconductor manufacturing where dimension of device is scaled down by a factor of in every eighteen months or so. It allows to pack more devices in similar area and thus adding more functionality in similar size. The process is named with the size of minimum channel length of the transistor it can offer. At present latest process node is 2nm which means transistor designed in this process can have minimum channel length as 2nm.

Dynamic QR Code for Further Reading

REFERENCES AND SUGGESTED READINGS

1. Lilienfeld, Julius Edgar (1926-10-08) "Method and apparatus for controlling electric currents" U.S. patent 1745175A
2. "Intel 45nm Hi-k Silicon Technology". Intel. Archived from the original on July 5, 2007.
3. Sedra, A. S. & Smith, K. C. (2004). *Microelectronic Circuits* (5th ed.). Oxford University Press. p. 250, Eq. 4.14
4. 1965 – "Moore's Law" Predicts the Future of Integrated Circuits". *Computer History Museum.*
5. Gray, P. R.; Hurst, P. J.; Lewis, S. H.; Meyer, R. G. (2001). *Analysis and design of analog integrated circuits* (4th ed.). New York: Wiley

❑❑

4 Power Devices

UNIT SPECIFICS

Through this unit we have discussed the following aspects:

- Introduction to power devices
- Operation of SCR, DIAC, TRIAC
- I-V characteristic of different power devices
- Application of power devices
- Comparison of switching behaviour of power devices

The practical applications of the topics are discussed for generating further curiosity and creativity as well as improving problem solving capacity.

Besides giving a large number of multiple choice questions as well as questions of short and long answer types marked in two categories following lower and higher order of Bloom's taxonomy, assignments through a number of numerical problems, a list of references and suggested readings are given in the unit so that one can go through them for practice. It is important to note that for getting more information on various topics of interest some QR codes have been provided in different sections which can be scanned for relevant supportive knowledge.

After the related practical, based on the content, there is a "Know More" section. This section has been carefully designed so that the supplementary information provided in this part becomes beneficial for the users of the book. This section mainly highlights the initial activity, examples of some interesting facts, analogy, history of the development of the subject focusing the salient observations and finding, timelines starting from the development of the concerned topics up to the recent time, applications of the subject matter for our day-to-day real life or/and industrial applications on variety of aspects, case study related to environmental, sustainability, social and ethical issues whichever applicable, and finally inquisitiveness and curiosity topics of the unit.

RATIONALE

This chapter provides a basic introduction to power electronics devices. It provides the details about operation, semiconductor physics and characteristic of different power electronics devices such as SCR, DIAC, TRIAC. Different modes of operations are considered to explain the working of these devices. It also provides details about characterizing these devices in laboratory.

PRE-REQUISITES

Mathematics: Algebra (Class XII)

Physics: Semiconductors (Class XII)

UNIT OUTCOMES

List of outcomes.of this unit is as follows:

U4-O1: Introduction to power devices

U4-O2: Operation of SCR, DIAC, TRIAC

U4-O3: I-V characteristic of different power devices

U4-O4: Application of power devices

U4-O5: Comparison of switching behaviour of power devices

Unit-4 Outcome	EXPECTED MAPPING WITH COURSE OUTCOMES (1 - Weak Correlation; 2 - Medium Correlation; 3 - Strong Correlation)				
	CO-1	CO-2	CO-3	CO-4	CO-5
U4-O1	3	3	3	-	3
U4-O2	3	1	2	2	1
U4-O3	3	1	3	1	2
U4-O4	1	–	3	1	2
U4-O5	1	3	3	-	3

4.1 INTRODUCTION

The discussion so far has been limited to the device operation, its semiconductor physics and to certain extent to its application. One important question we should ask at this point is what is the permissible range of voltages and currents which can be applied safely to these devices without affecting its operations? The answer to this depends upon the geometry and semiconductor physics of electronic device in consideration. The geometry and semiconductor physics of an electronic device limits the maximum voltage or current which can be applied safely to it without affecting its operation. For example if a two terminal device has a fixed spacing between its two terminal then an increasing voltage across its terminal results in an increasing electric field. If electric field increases beyond a value then it results in impact ionization or avalanche which if becomes irreversible can damage the device. Similarly increasing current between its terminals increases thermal losses in device which increases its temperature. If the temperature of device increases beyond a point then it may affect its operation and eventually damage the device. These thermal losses depends upon the internal resistance of the device which is a function of its cross sectional area. A larger cross sectional area is required to achieve a smaller internal resistance. Thus a relatively larger cross sectional area is required to accommodate a large current as it results in smaller internal resistance.

Thus, the maximum value of each operating point (voltage or current or power) in a device depends on its semiconductor physics and geometry. These operating points are important device specifications which are defined as rated voltage or rated current or rated power of the device. If a device is a multi-terminal device then many such rated voltage are defined across every possible pair of terminals and many such rated currents are defined between every pair of terminals. A rated voltage is the maximum value of voltage difference which can be applied across its pair of terminals safely without affecting its operation or damaging the device.

Semiconductor devices which operates at supply voltages ranging from 1-12V or currents ranging up to few hundreds of mA or at max to several of Amperes are generally termed as electronic devices. However, in certain scenarios interest might be in controlling a much higher supply voltages say ranging up to kV or current ranging up to kA or power ranging in kW or MW. In such scenarios, nominal electronic devices may not work as it has limited current or voltage or power handling capacity. The devices which are designed to operate at higher value of voltages or currents are termed as power electronic devices. In this chapter we will study several of such power electronic devices.

Power electronic devices can be broadly classified into three categories.

1. **Uncontrolled Devices:** These devices do not have any external control on its on and off operation. It's on and off operation is defined by the device itself. Example of such devices are power diode and DIAC
2. **Semi controlled Devices:** These devices uses an external control in triggering either it's on state or its off state but not both. The other state is triggered based on device operation and its physics. Example of such devices are Silicon controlled rectifier (SCR).
3. **Fully controlled Devices:** These devices have an external control over its on and off operation. Thus these devices can be fully controlled using this external input. Example of such devices are Power BJT, Power MOSFET, IGBT.

4.2 POWER DIODES

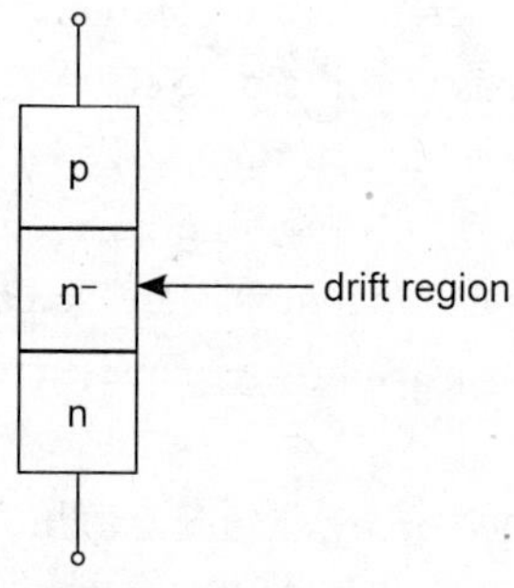

Fig. 4.1: Semiconductor structure of a power diode

Diodes used in high power applications are known as power diode. Power diode has an additional drift layer between its p-n junction as shown in Fig. 4.1 to accommodate an higher electric filed so that it can be operated at higher voltages. This additional layer is generally a lightly doped n-layer which is almost like an intrinsic semiconductor and hence sometime these diodes are also called as PIN diode. Its V-I characteristic is similar to conventional P-N junction diode with a difference that its reverse breakdown voltage is much higher due to additional drift region and its forward characteristic is almost linear due to additional resistance caused by lightly doped n- region.

4.3 SILICON CONTROLLED RECTIFIER (SCR)

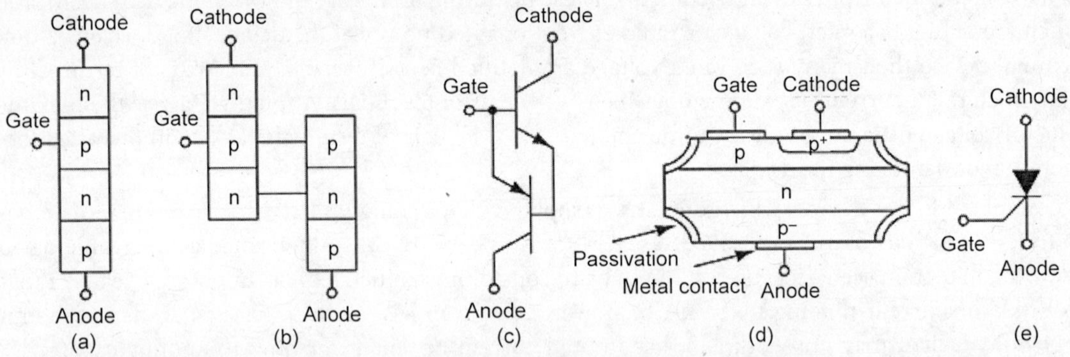

Fig. 4.2: (a) Block diagram of a SCR (b) Splitting of 4-layer SCR into two three layer semiconductor device(BJT) (d) representation of a SCR device as two connected BJT transistors (d) schematic of a SCR device (e) Symbol of a SCR

Thyristor is a group of semiconductor device which have four layer structure made from alternating layers of p-type and n-type semiconductor materials. Silicon controlled rectifier (SCR) is a power device belonging to this family. The block diagram of an SCR is shown in Fig.4.2(a). It can be viewed as an interconnection of two BJTs as shown in Fig.4.2(b). Its equivalent circuit is shown in Fig.4.2(c). In order to understand its operation, we use the equivalent circuit shown in Fig.4.2(c). A positive potential difference is needed between its anode(A) and cathode (K) to switch the device on. However, even after applying a positive potential difference between its anode and cathode, the device conducts only if its gate current increases beyond a value known as latching current. If initially SCR is off then its gate current is zero. The gate current is increased by applying a positive potential to its gate with respect to its cathode terminal. This nonzero gate current which is also the base current of NPN transistor in equivalent circuit shown in Fig.4.2(c) generates a much larger collector current for this NPN transistor. The collector current of NPN transistor is also the base current of PNP transistor in equivalent circuit. This large base current of PNP transistor turns it on and generates a large collector current for it. The large collector current of PNP transistor becomes the base current of NPN transistor and thus this becomes a latching action. Thus if the gate current is increased above a certain value known as latching current then latching action results in a sufficiently large base current which may keep device on even if external excitation is removed.

Thus to summarize a SCR is switched on by ensuring following conditions (a) Its anode potential is higher than its cathode potential. (b) A gate pulse has been provided to its gate terminal such that $V_{GK} > 0$. (c) Its gate current becomes higher than latching current.

4.3.1 Triggering Circuit of an SCR

The process of switching on the SCR by applying a pulse at its gate terminal is known as triggering of SCR. An example circuit to trigger the SCR is shown in Fig.4.3. This trigger circuit is an electronic circuit operating with a lower supply V_{cc} which might be 5V or lower while the SCR circuit might be operating at much higher supply voltage ranging up to few kV. Hence, the two circuits must be isolated which is done here using transformer. A pulse when applied to a transformer may saturate transformer if it is on for a longer duration. However such a pulse might be needed to initiate latching action in SCR. Hence, a modified pulse is generated by performing AND operation of long pulse with a high frequency clock. The modified pulse allows transformer to de-flux before the flux increases again. The output of AND gate may have limited current driving capability and thus to increase its driving capability a common emitter amplifier is used. The collector current of common emitter amplifier drives the lower end of primary coil of the transformer while the higher end is connected to power supply of electronic circuit. The higher end of secondary coil of transformer is connected to the gate of thyristor through a diode D_1 which ensures that only positive voltage is applied to gate terminal with respect to its cathode terminal. Its lower end is connected to the ground of power electronic circuit.

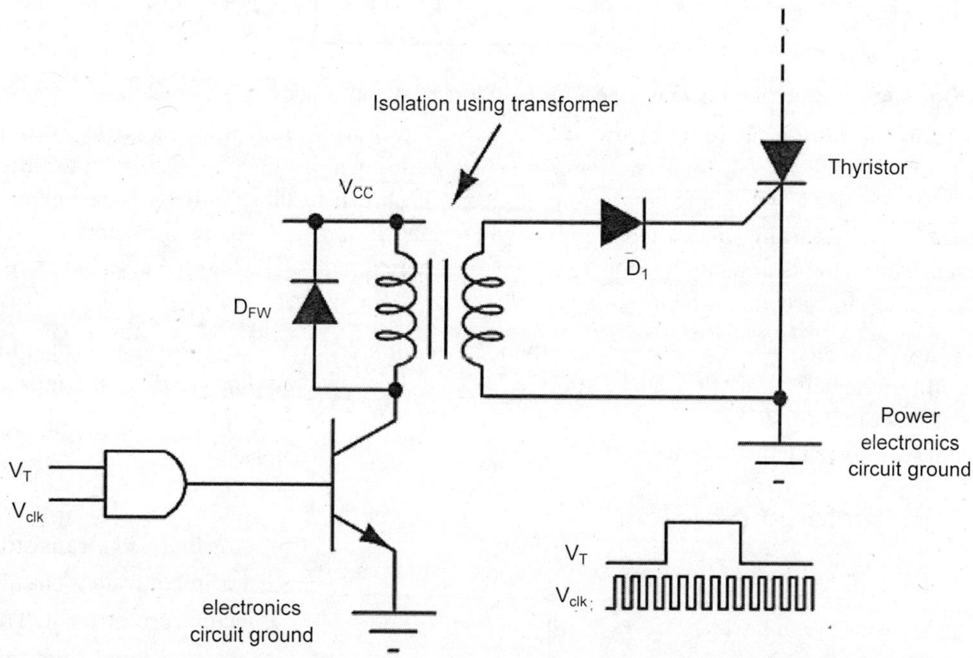

4.3.2 Commutation Circuit of an SCR

Now to turn off a SCR one need to reduce the current flowing between its anode and cathode terminal below a certain value known as holding current. However, even after reducing the current flowing through thyristor to zero value or below holding current there will be additional charge carriers within thyristor. So if a large negative voltage is applied between anode and cathode terminal then these charges get recombined much faster and thyristor can achieve its forward blocking state. The circuits used to perform this task is known as commutation circuit. There are two categories of commutation in SCR.

1. **Natural commutation:** This type of commutation is possible when input source is AC. As for an AC signal the current automatically becomes zero and it experiences a negative voltage across it because polarity of signal source changes once current becomes zero. Hence SCR turns off automatically.

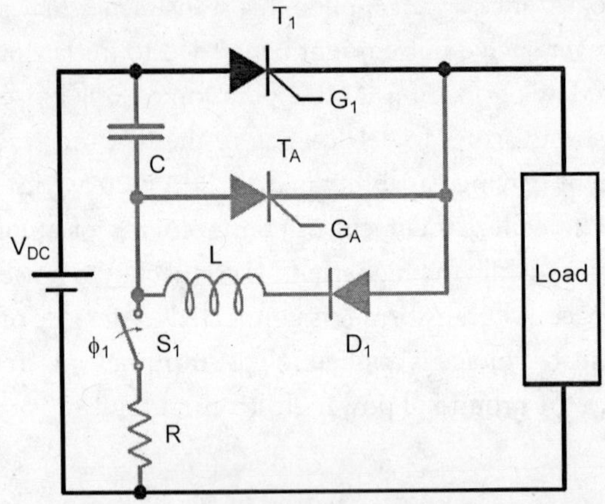

Fig. 4.4: Commutation circuit for performing forced commutation of a thyristor

2. **Forced commutation:** If the input source is DC source then current may not necessarily become zero with time hence an additional mechanism is needed to turn off the SCR. One such example circuit is shown in Fig.4.4. Here, commutation circuit is shown in blue. Initially capacitor C is charged to V_{DC} by turning on switch S_1. Now when SCR T_1 is triggered, current start flowing through it. The currents flowing through T_1 takes two different paths. The first path is from source $V_{DC} - T_1 -$ Load and second path is through $C - T_1 - D_1 - L$. The second path $C - T_1 - D_1 - L$ is also a resonating path. Hence it changes the polarity of capacitor voltage within $T = \pi\sqrt{LC}$. This time correspond to the half cycle of resonating period of this path. Now if auxiliary SCR T_A is triggered exactly at this instant then it will apply a reverse voltage across SCR T_1 and thus bringing current through it to zero value much quickly.

4.3.3 I-V Characteristic of SCR

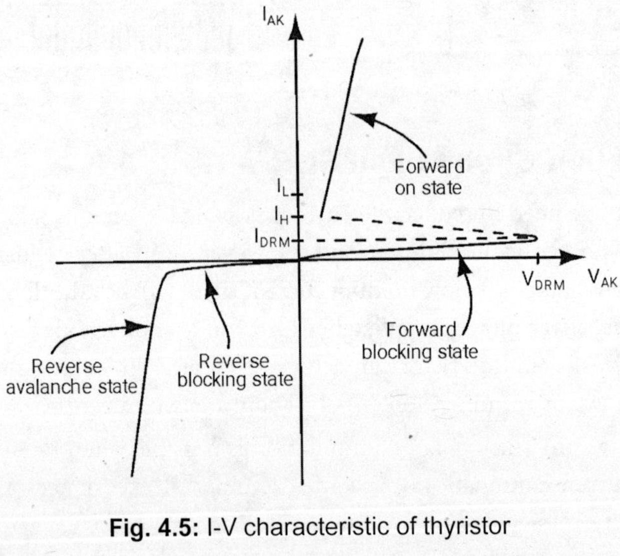

Fig. 4.5: I-V characteristic of thyristor

Fig.4.5 shows the I-V characteristic of SCR. If SCR is initially in its off state which is also known as forward blocking state then with increasing voltage VAK its current increases slightly and it offers a very high resistance. However, if a trigger pulse is applied at its gate terminal to trigger the SCR then it moves to forward on state where its current increases much quickly and voltage VAK across its anode and cathode terminal reduces substantially. In this state it offers a much lower resistance and thus behaves like an on switch. If current through SCR increases beyond a value known as latching current IL then it remains in forward on state even if gate trigger pulse is removed. Once the SCR is latched it can be turned off only if the current through it becomes smaller than IH, where IH is termed as holding current. If the voltage V_{AK} across SCR becomes negative then it carries a much smaller current due to minority carriers. This region of I-V characteristic is known as reverse blocking state. However when VAK reaches to sufficiently large negative value the current through it again becomes substantially high due to avalanche breakdown happening in its semiconductor junction. This region of I-V characteristic of SCR is known as reverse avalanche state.

4.3.4 Types of SCR

Based on its uses SCR can be divided into following two categories.

1. **Converter grade SCR:** These thyristors mostly used for rectification or conversion of AC into DC. These SCRs have higher current, voltage and power rating than inverter grade SCRs.
2. **Inverter grade SCRs:** These SCRs are used in conversion of DC into AC signal. It has a faster turn on and turn off time to allow the switching at faster speed. These thyristors have relatively smaller current, voltage and power rating than converter grade SCRs as it need to be switched on and off faster.

From the discussion above we can see that SCRs are quite helpful in controlling the input signal. However it has a unidirectional control where it can only control the positive half cycle of an AC signal. Thus for an effective bidirectional control of an AC signal, we need a power device which should be able to conduct in both direction so that it can be designed to have a bidirectional control.

4.4 DIODE FOR ALTERNATING CURRENT (DIAC)

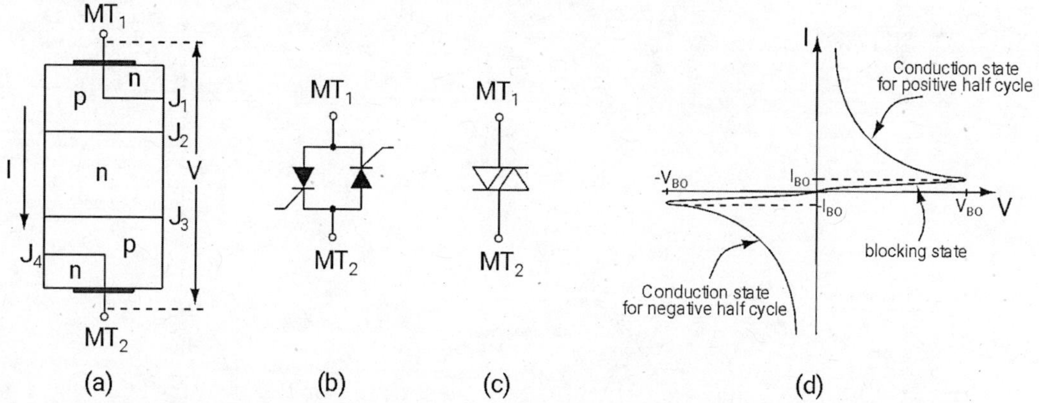

Fig. 4.6: (a) Device structure of a DIAC (b) Equivalent circuit of DIAC using SCR where gate of both SCRs are left open (c) Symbol of a DIAC (d) I-V characteristic of DIAC

DIAC is a bidirectional power electronics device which is abbreviation of Diode for Alternating Current. Its device structure is shown in Fig.4.6(a). It can be modelled as two SCR connected in anti-parallel fashion as shown in Fig.4.6(b) where gate terminal of SCR is left open. The device functions as follows. Assume that device is initially off and a positive terminal of supply is connected to MT_1 and a negative terminal of supply is connected to MT_2 so that voltage across the device V is positive. Now as V increases I also increases but with a very small slope as the device is largely off due to reverse bias in p-n junction (J_3). This region of operation of device is known as blocking state. Now if V increases beyond a value V_{BO} which is the breakover voltage of SCR then concentration of holes in its middle layers increases significantly changing the nature of semiconductor and thus making it a continuous p-type semiconductor between MT_1 and MT_2 terminals. This reduces the resistance of device and hence current through it increases while voltage across it decreases. This region of operation is known as on state of the DIAC. The device has a symmetrical structure and hence it can be switched to on state when voltage across it becomes larger than breakover voltage in any direction. The device moves back to its off state only when the current through it becomes smaller than breakover current or holding current. Its symbol is shown in Fig.4.6(c) and its I-V characteristic is shown in Fig.4.6(d). It has a symmetrical I-V characteristic which is self-explainable as the device is symmetrical.

4.5 TRIODE FOR ALTERNATING CURRENT (TRIAC)

TRIAC is another power electronic device which is an abbreviation for TRIode for Alternating Current. It is a bidirectional power electronics device which can be used to control the current in both directions. The Fig.4.7(a) shows the device structure of a TRIAC. It can also be viewed as a combination of two SCRs connected in anti-parallel fashion with its gate terminal shorted as shown in Fig.4.7(b). The Fig.4.7(c) shows the symbol of TRIAC. The I-V characteristic of TRIAC is shown in Fig.4.7(d) where potential difference between its terminal G and MT_1 is positive.

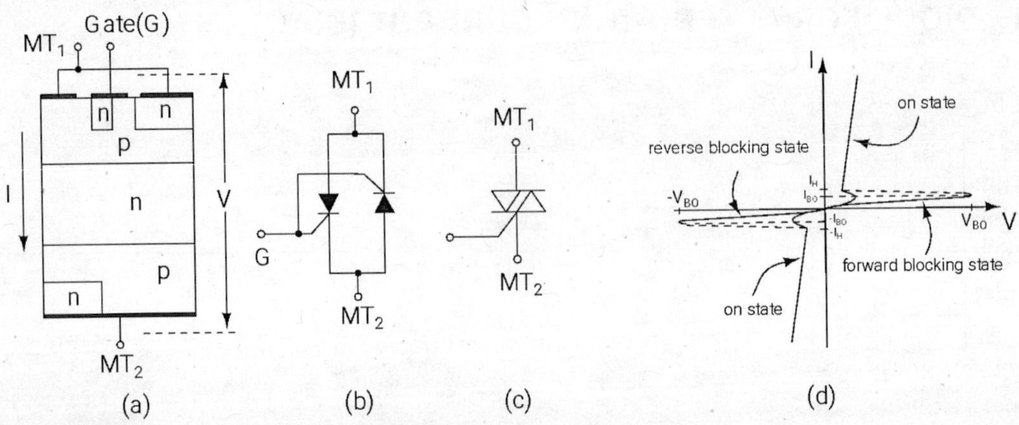

Fig.4.7(a) Device structure of a TRIAC (b) Equivalent circuit of TRIAC using SCR where gate of both SCRs are left open (c) Symbol of a TRIAC (d) I-V characteristic of TRIAC when gate to MT_1 potential is positive

In order to understand the working of TRIAC, we need to consider the potential of MT_2 and G terminal of TRIAC with respect to MT_1. There are four possible combination of polarity of these potentials. Each of these combination defines one mode of TRIAC operation which in its V-I characteristic can be defined as quadrant. The conditions for operating in these modes are shown in Fig.4.8(a)-(d). Its equivalent circuits are shown in Fig4.8(e)-(h).

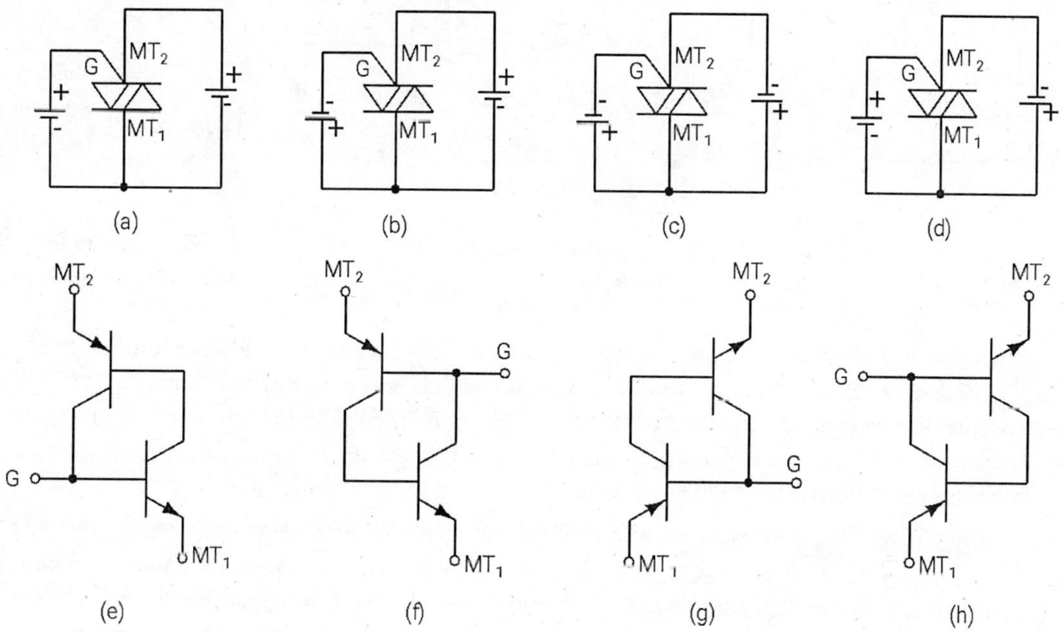

Fig. 4.8: (a) Condition for TRIAC operation in quadrant I of I-V characteristic (b) Condition for TRIAC operation in quadrant II of I-V characteristic (c) Condition for TRIAC operation in quadrant III of I-V characteristic (d) Condition for TRIAC operation in quadrant IV of I-V characteristic (e) Equivalent circuit of TRIAC operating in quadrant I (f) Equivalent circuit of TRIAC operating in quadrant II (g) Equivalent circuit of TRIAC operating in quadrant III (h) Equivalent circuit of TRIAC operating in quadrant IV

In quadrant I and II, MT_2 is at higher potential with respect to MT_1 and thus current in TRIAC flows from MT_2 to MT_1. The path taken by current from MT_2 to MT_1 passes through layer P-N-P-N. The N region connected to MT_2 does not play any significant role in this case. In quadrant III and IV, MT_2 is at lower potential with respect to MT_1 and thus current in TRIAC flows from MT_1 to MT_2. The path taken by current from MT_1 to MT_2 passes through layer P-N-P-N. The N region connected to MT_1 does not play any significant role in this case and is only active during initial triggering.

The current flow in TRIAC in quadrant I operation is shown in Fig.4.9(a). In quadrant 1 potential of G terminal is higher than MT_1 and hence NPN transistor gets activated which in turn supplies current $I_{c,n}$ to base of PNP transistor and thus creating a path from MT_2 to MT_1 for current to flow as shown in Fig.4.9(a). This operation is quite similar to the SCR operation with a difference that there exist a ohmic path from gate terminal to MT_1 region through P type semiconductor shown by $I_{b,r}$ in Fig. 4.9 (a). This requires gate current drive to be higher than in the case of SCR.

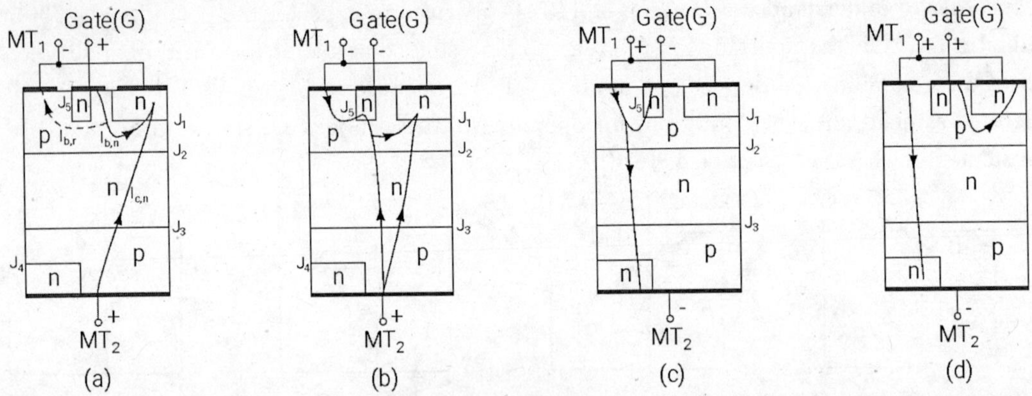

Fig. 4.9: (a) Current flow in TRIAC in quadrant I operation (b) Current flow in TRIAC in quadrant II operation (c) Current flow in TRIAC in quadrant III operation (d) Current flow in TRIAC in quadrant IV operation

The current flow in TRIAC in quadrant II operation is shown in Fig.4.9(b). In quadrant 2 potential of G terminal is negative with respect to MT_1 however MT_2 is still positive with respect to MT_1. In this case initially the p-n junction (J_5) formed due to p region under MT_1 and n region under gate becomes forward biased and conducts a current. This current then triggers the NPN transistor formed by (J_1 and J_2) which in turn triggers the PNP transistor formed by (J_2 and J_3) allowing current flow from MT_1 to MT_2. The current flow is same as in the case of quadrant 1 operation.

The current flow in TRIAC in quadrant III operation is shown in Fig.4.9(c). In quadrant 3 potential of G terminal is negative with respect to MT_1 however MT_2 is also now negative with respect to MT_1. Here initially the p-n junction (J_5) formed due to p region under MT_1 and n region under gate becomes forward biased and conducts a current. This injects minority carriers in p region, some of which cross over J_2 and enters into n region below it and lowers its potential thus activating PNP transistor formed by J_2 and J_3. This further activates the NPN transistor formed by J_3 and J_4 which in a way allows a current flow between MT_1 and MT_2.

The current flow in TRIAC in quadrant IV operation is shown in Fig.4.9(d). In quadrant 4 potential of G terminal is positive with respect to MT_1 however MT_2 is now negative with respect to MT_1. Here initially the p-n junction (J_1) formed due to p region under gate and n region under MT_1 becomes forward biased and conducts a current. This injects minority carriers in p region, some of which cross over J_2 and lowers its potential. This activates the PNP transistor formed by J_2 and J_3 which then activates the NPN transistor formed by J_3 and J_4. Thus a current start flowing between MT_1 and MT_2.

4.6 APPLICATION OF SCR, DIAC AND TRIAC

SCRs are used as a switch to control the flow of current through load such as DC motor to control its speed. However, SCRs can perform this control operation only if its anode potential is higher than its cathode potential. This kind of control is termed as unidirectional control. The Fig.4.10 shows the use of SCR T_1 to control the speed of DC motor. Here only when T_1 is triggered, current start flowing to DC motor. The circuitry shown in blue is for commuting the main SCR T_1 and its operation has already been described earlier in this chapter.

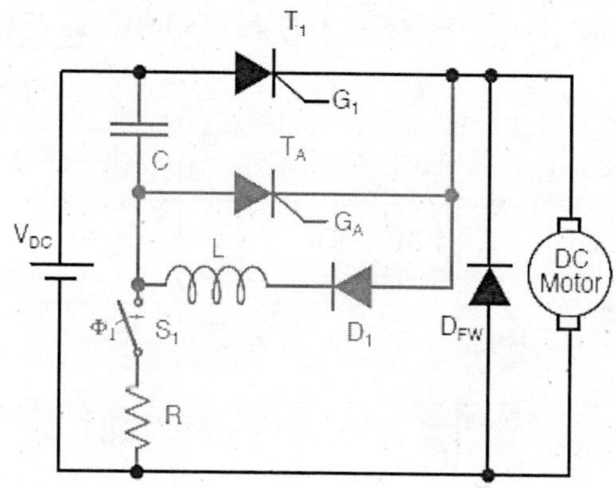

Fig. 4.10: Speed control of DC Motor using SCR

DIAC and TRIAC are bidirectional switch where DIAC is a bidirectional switch without any external control and TRIAC is a bidirectional switch with an external control. These switches are useful in controlling application such as dimming of light or speed control of Motors.

The Fig.4.11 shows the use of DIAC and TRIAC in controlling the intensity of Light bulb. Here TRIAC is used as a main switch while DIAC is used as a switch to trigger the TRIAC. Initially when voltage is zero, TRIAC T_1 as well as DIAC D_1 are off. Now as voltage increases. It charges the capacitor C which increases the voltage across DIAC. The DIAC triggers on as soon as voltage across reaches above its breakover voltage. This allows DIAC to conduct and it triggers the TRIAC which then allow the current to flow through the bulb. The TRIAC gets triggered off as soon as voltage VAC again becomes zero. The operation of this circuit for negative half cycle of AC voltage is similar to the positive half cycle. Thus this circuit controls the current flowing through electric bulb and hence controls its intensity. The tenability of voltage across DIAC is achieved by controlling the value of resistance R which is tuneable in this case.

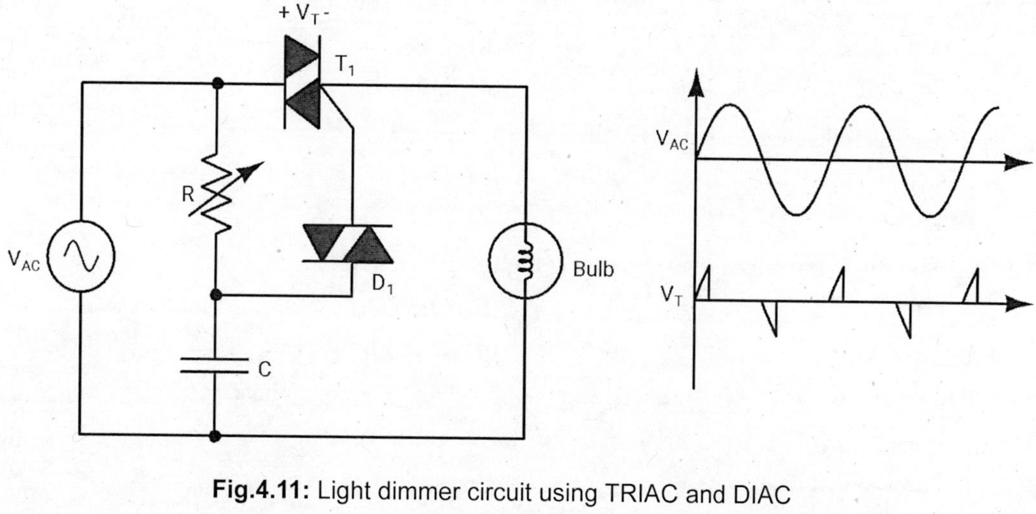

Fig.4.11: Light dimmer circuit using TRIAC and DIAC

4.7 COMPARISON OF DIFFERENT SWITCHING ELEMENTS

S.No.	SCR	DIAC	TRIAC	MOSFET
Function	Rectification with a control input	Bidirectional switch without external control input	Bidirectional control switch	Switching as well as amplification
Direction of current	Unidirectional	Bidirectional	Bidirectional	Bidirectional
Controlled device	Yes	No	Yes	Yes
Power Rating	Highest	High	Medium	Medium
Switching frequency	50-500Hz	50-500Hz	50-500Hz	100Hz-100kHz

UNIT SUMMARY

This unit provides an introduction of power electronic semiconductor devices including DIAC,TRIAC, Power diodes, SCR and Power MOSFET. It discusses the device operation and characteristics. The difference between characteristics of different devices are discussed and analyzed. Applications using these power semiconductor devices are also discussed.

EXERCISES

Multiple Choice Questions

1. Which of the following device does not have an external control?
   (a) SCR (b) DIAC
   (c) TRIAC (d) MOSFET
2. which of the following device has highest power rating?
   (a) SCR (b) DIAC
   (c) TRIAC (d) MOSFET
3. Which of the following can be operated at highest frequency?
   (a) SCR (b) DIAC
   (c) TRIAC (d) MOSFET
4. Fill in the blanks. A DIAC conducts when applied voltage across it becomes higher to ___
   (a) Breakover voltage (b) cut-in voltage
   (c) Threshold voltage

5. Which of the following component can perform wave shaping of a sinusoidal signal with zero bias
   (a) Diode (b) Triode
   (c) DIAC (d) TRIAC
6. Which of the following device allows bidirectional current flow
   (a) BJT (b) MOSFET
   (c) IGBT (d) TRIAC
7. Which of the following statement is correct?
   (a) TRIAC is a unidirectional device
   (b) TRIAC is a bidirectional device
   (c) TRIAC is equivalent to two SCRs connected in parallel
   (d) DIAC is a three terminal device.
8. A DIAC is a ___ layer device. Fill in the blanks with one of the following option whichever is correct.
   (a) 2 (b) 3
   (c) 4 (d) 5

ANSWERS															
1.	(b)	2.	(a)	3.	(d)	4.	(a)	5.	(d)	6.	(d)	7.	(b)	8.	(c)

Short Answer Type Questions

1. What is the difference between power devices and conventional semiconductor devices?
2. Discuss about the different category of power electronic devices.
3. What is the purpose of drift layer in power electronic devices?
4. What is the purpose of trigger in SCR?
5. What is holding current in SCR? Define it.
6. Discuss about the different category of SCR.
7. What is latching current in an SCR?

Long Answer Type Questions

1. What is DIAC? Show its device structure. Explain its working.
2. What is SCR? Show its device structure. Explain its working.
3. What is TRIAC? Show its device structure. Explain its working.
4. Describe about different quadrant of TRIAC operation? Which quadrant is more common to be used in its operation?
5. Describe the applications of SCR and TRIAC?
6. How SCRs are triggered? Give examples of some trigger circuits.

Numerical Problems

1. The circuit shown in Fig.4.12 uses a DIAC to trigger the TRIAC. If switch is closed then calculate the current through the diode.

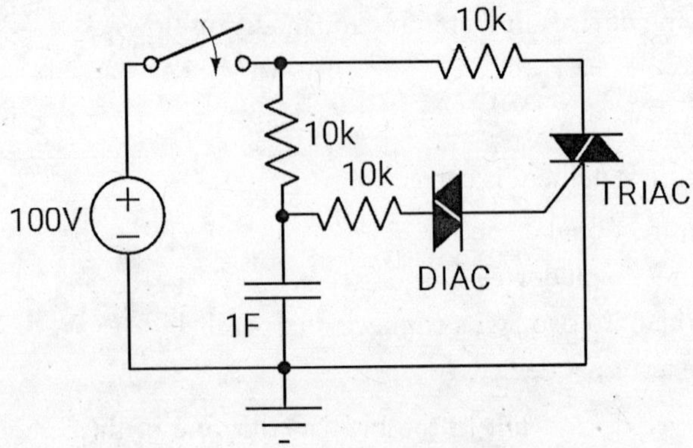

Fig. 4.12: Circuit for problem Q.N4.1

2. An SCR is used to rectify the AC sinusoidal signal of 240V which has a frequency of 50Hz. If the blocking voltage of SCR is 100V then calculate the percentage time during one cycle of input signal for which SCR remains off.
3. The circuit shown in Fig.4.13 uses an SCR S_1. Find its gate triggering current If the circuit gets triggered when input source becomes equal to 10.7V. Also find the holding current if SCR turns off when input source becomes equal to 2.7V. The junction voltage of p-n junction is assumed as 0.7V.

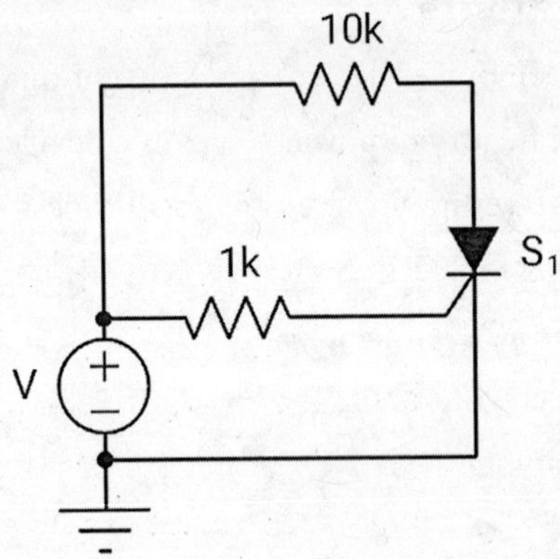

Fig. 4.13: Circuit for problem Q.N4.3

PRACTICAL

Experiment-1: Study of characteristic of DIAC

Aim: Following are the aim of this experiment

1. Finding V-I characteristic of DIAC.
2. **Equipment and Components:**

S.No.	Name of the component	Specification	Quantity
1	DIAC	NTE6407	1
2	Resistor	1kΩ	1
3	Resistor	10kΩ	1
5	Power supply	0-30V	1
6	Ammeter		1
7	Voltmeter		1
8	Wires		

Procedure:

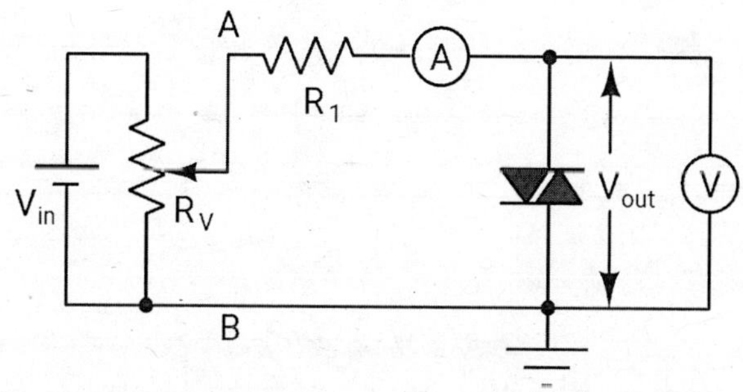

Fig. 4.14: Circuit diagram for characterizing DIAC

1. This experiment requires a variable power supply. One possible way to obtain a variable power supply is to use a DC source along with a variable resistor as shown in Fig.4.14.
2. Note down the resolution and any error reading on Ammeter and voltmeter.
3. Arrange the components and make all necessary connection to get the circuit shown in Fig.4.14.
4. Check the reading on ammeter, it should be close to zero.
5. Now start increasing the voltage output of power supply in steps of 1V by tuning the variable resistor. Tabulate the readings of voltmeter and ammeter in Table.1
6. As the voltage across DIAC reaches to breakover voltage the current through it start increasing sharply and voltage across it reduces substantially hence reduce your step size in this region. You will notice that current will change linearly
7. Now change the polarity of signal source to bias DIAC with opposite polarity and repeat the steps 4-6 above.

Observation

S.No.	Voltage	Current

Experiment-2: Plot the characteristic of a TRIAC.

Aim:

3. To plot the characteristic of a TRIAC

Equipment and Components:

S.No.	Name of the component	Specification	Quantity
1	TRIAC	BT136	1
2	Resistor	100Ω	2
3	Resistor	1kΩ	1
5	Power supply	0-30V	2
6	Ammeter		2
7	Voltmeter		1
8	Wires		

Procedure:

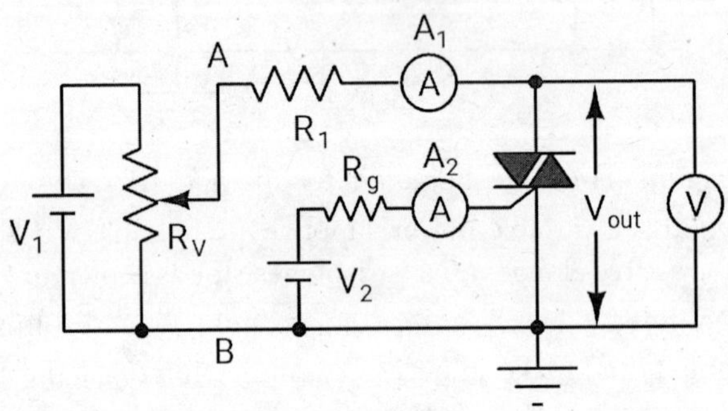

Fig. 4.15: Circuit diagram for characterizing the TRIAC

1. Check the power supply and all the components.
2. Note down the resolution and any error reading on Ammeter and voltmeter.
3. Arrange the components and make all necessary connection to get the circuit shown in Fig.4.15.
4. Check the reading on ammeter, it should be close to zero.
5. Initially keep the gate current 0. Now start increasing the voltage output of power supply V_1 in steps of 1 V. Tabulate the readings of voltmeter and ammeter A_1 in Table.1
6. Now increase the gate voltage slowly until the TRIAC gets triggered. Note down the reading of current in Ammeter A_2. Also note down the reading of voltage in voltmeter and Ammeter A_1.
7. Now reduce gate voltage to zero and also start reducing V_1. The reading of ammeter reduces slowly but at certain point it quickly drops to zero. The last reading of ammeter before it quickly drops to zero is the reading of holding current.
8. To characterize TRIAC in other modes. Polarity of V_1 and V_2 is changed and steps 5-7 are repeated.
9. Plot the reading on same graph.

KNOW MORE

Silicon based power semiconductor devices are now reaching to its limitation where its performance is getting limited due to restricted blocking voltage, limited junction temperature capability, limited efficiency and thermal capabilities. Wide bandgap semiconductor material is a class of materials whose bandgap is higher than bandgap of silicon. These materials offer some advantages when compared to silicon. For example Silicon carbide (SiC) which is a wide bandgap material allows devices to have a higher blocking voltage. It has been demonstrated that switches designed using SiC can operate at 650V or higher. Another wide bandgap material Gallium nitride (GaN) has a much smaller gate charge and a low threshold voltage even at higher frequency, and thus allowing the devices to operate at much higher frequency with a better efficiency.

Dynamic QR Code for Further Reading

REFERENCES AND SUGGESTED READINGS

1. Owen, Edward L. "SCR is 50 years old [history]." *IEEE Industry Applications Magazine* 13.6 (2007): 6-10.
2. Yoshikawa, A. (2007). "Development and Applications of Wide Bandgap Semiconductors". In Yoshikawa, A.; Matsunami, H.; Nanishi, Y. (eds.). *Wide Bandgap Semiconductors*. Springer.
3. Semiconductor, O. N. "Thyristor theory and design considerations." *HBD855/D* (2006): 11.
4. Allan, Roger. "Power semiconductors." *IEEE spectrum* 12.11 (1975): 37-45.
5. Greenburg, Ralph. "Consumer applications of power semiconductors." *Proceedings of the IEEE* 55.8 (1967): 1426-1434.
6. Herrick, Phillip R., and E. Keith Howell. "Solid-State Lighting Control." *IEEE Transactions on Industry and General Applications* 3 (1965): 225-232.

5 Amplifier and Oscillator

UNIT SPECIFICS

Through this unit we have discussed the following aspects:

- Introduction to amplifiers and its design specifications
- Derivation of two port parameter of an amplifier
- Introduction to feedback, positive and negative feedback
- Different feedback amplifier configuration
- Introduction to oscillators, different types of oscillators

The practical applications of the topics are discussed for generating further curiosity and creativity as well as improving problem solving capacity.

Besides giving a large number of multiple choice questions as well as questions of short and long answer types marked in two categories following lower and higher order of Bloom's taxonomy, assignments through a number of numerical problems, a list of references and suggested readings are given in the unit so that one can go through them for practice. It is important to note that for getting more information on various topics of interest some QR codes have been provided in different sections which can be scanned for relevant supportive knowledge.

After the related practical, based on the content, there is a "Know More" section. This section has been carefully designed so that the supplementary information provided in this part becomes beneficial for the users of the book. This section mainly highlights the initial activity, examples of some interesting facts, analogy, history of the development of the subject focusing the salient observations and finding, timelines starting from the development of the concerned topics up to the recent time, applications of the subject matter for our day-to-day real life or/and industrial applications on variety of aspects, case study related to environmental, sustainability, social and ethical issues whichever applicable, and finally inquisitiveness and curiosity topics of the unit.

RATIONALE

This chapter provides an introduction to amplifier and its design specification. It provides the basic definition of different design specification. It then derives the two port parameters of an amplifier emphasizing the requirements on device for designing an amplifier. The chapter then introduces the concept of feedback and provides the details about different types of feedback such as positive and negative feedback. Different feedback amplifier configurations are discussed in detail. It also provides an introduction to oscillators and different types of oscillators providing an insight into its design details.

PRE-REQUISITES

Mathematics: Algebra (Class XII)
Physics: Semiconductors (Class XII)

UNIT OUTCOMES

List of outcomes of this unit is as follows:

U5-O1: Introduction to amplifiers and its design specifications
U5-O2: Derivation of two port parameter of an amplifier
U5-O3: Introduction to feedback, positive and negative feedback
U5-O4: Different feedback amplifier configuration
U5-O5: Introduction to oscillators, different types of oscillators

Unit-5 Outcome	EXPECTED MAPPING WITH COURSE OUTCOMES (1 - Weak Correlation; 2 - Medium Correlation; 3 - Strong Correlation)				
	CO-1	CO-2	CO-3	CO-4	CO-5
U5-O1	1	2	3	3	3
U5-O2	2	2	2	2	3
U5-O3	2	2	3	1	2
U5-O4	1	3	3	1	2
U5-O5	1	3	3	3	3

5.1 INTRODUCTION

When a signal is processed, disturbance from other sources or from device itself start affecting the quality of signal in consideration. These disturbances which are unwanted in nature are termed as noise. The metric which is used to define the quality of signal is signal to noise ratio (SNR). It is defined as a ratio of signal power to noise power, which is mostly defined in decibels using the expression given in (5.1)

$$SNR(in\ dB) = 10\log_{10}\frac{\text{Signal Power}}{\text{Noise Power}} \tag{5.1}$$

If SNR is higher than 0 dB then only signal is distinguishable from noise otherwise noise overpower the signal and it is not easy to process signal in such case. Hence, if signal is weak then it is advisable to increase its power before transmitting it through a noisy channel or medium. The process of increasing signal amplitude or power is known as amplification. An amplifier is a circuit block designed to perform this task. In previous chapters we have seen some examples of amplifier circuits. In this chapter we will develop a systematic approach to analyze and design amplifiers.

5.2 DEFINITIONS OF AMPLIFIER DESIGN SPECIFICATIONS

The design of an amplifier begins with its design specifications. The design specifications of an amplifier are set of amplifier characteristics such as gain and bandwidth which depends on its desired applications. Following are few important design specifications of an amplifier,

1. **Gain:** Gain is defined as the ratio of output signal amplitude to input signal amplitude. For an electrical signal, input or output can either be voltage or current and hence there are four possibilities as listed in Table.5.1.

Table 5.1: Categories of amplifiers

	Input	Output	Type of amplifier	Unit
1	Voltage signal	Voltage signal	Voltage amplifier	Unitless
2	Voltage signal	Current signal	Transadmittance amplifier	Mho
3	Current signal	Voltage signal	Transimpedance	Ohm
4	Current signal	Current signal	Current amplifier	Unitless

   If both input and output are voltage signal then amplifier is voltage amplifier. Its gain is a unitless quantity as it is simply a ratio. If both input and output are current signals then amplifier is a current amplifier. Its gain is also a unitless quantity.

   If input is a current signal and output is a voltage signal then amplifier is termed as a transimpedance amplifier. Its gain has a unit of impedance. If input is a voltage signal and output is a current signal then amplifier is termed as a transadmittance amplifier. Its unit is Mho or Siemens.

2. **Bandwidth:** Bandwidth is another important amplifier specification. It is defined as a range of frequencies where gain of the amplifier is above certain range of amplitude. Generally there are two definition of bandwidth. A 3-dB bandwidth is a range of frequencies within which gain does not fall more than 3 dB of peak gain. Sometime another different definition of bandwidth known as unity gain bandwidth is used. The UGB of a circuit is a range of frequencies within which gain of the system remains above unity.

3. **Input impedance:** Input impedance decides the coupling of signal to the amplifier. If input signal is a voltage signal generated by a real voltage source which can have any series impedance then it couples completely to amplifier only if input impedance of amplifier is infinity. Hence input impedance of a voltage input amplifier should be infinite or very large. Similarly if input signal is a current signal generated by a real current source having finite shunt resistance then it couples completely to amplifier only if input impedance of amplifier is zero. Hence input impedance of a current input amplifier should be zero or very low.

4. **Output impedance:** Output impedance defines the coupling of output signal of amplifier to the load. If amplifier is a voltage output amplifier then output impedance of amplifier should be zero or low to make it behave like an ideal voltage source. Hence output impedance of a voltage output amplifier should be zero or very low. Similarly if amplifier is a current output amplifier then its output impedance should be infinite to make it behave like an ideal current source. Hence output impedance of a current output amplifier should be infinite or very high.

5. **Swing limits :** It defines how high or low a signal can swing without taking any device out of its intended region of operation. Swing limits at input of amplifier is known as input swing limits and swing limits at output node of amplifier is known as output swing limits.
6. **Noise:** All real circuit designed using real devices introduce several different sources of noise. The analysis of noise is simplified by referring all these noise sources at input in a form of input referred voltage noise and input referred current noise. These noise sources together define a SNR at the input of amplifier. This SNR remains same even at the output of amplifier as signal and input referred noise see the same gain from input to output. Thus input referred voltage noise and input referred current noise are another important specification of amplifier.
7. **Power dissipation:** Power dissipation is an important metric where objective is always to minimize it. It becomes more important in case of wireless and portable device where power available is limited and hence designing a low power circuit becomes a necessity.
8. **Nonlinearity:** Although designer's intention is always to design a linear amplifier. But as we know now that amplifiers are designed using nonlinear devices. These amplifier circuits are operated in small signal region of these nonlinear devices where they behave like a linear device. The linearity performance of these amplifiers are characterized using several different parameters such as HD2, HD3, THD and SFDR.

   HD2 is the ratio of the strength of second order harmonic to fundamental signal component. It is generally measured in dB as expressed in (5.2).

$$HD_2(dB) = 10\log_{10}\left(\frac{V_2}{V_1}\right) \tag{5.2}$$

   Here V_2 defines the rms value of 2nd harmonic in volts and V_1 defines the rms value of fundamental component of signal in volts.

   Similarly, HD3 is the ratio of the strength of third order harmonic to fundamental signal component. It is generally measured in dB as expressed in (5.3).

$$HD_3(dB) = 10\log_{10}\left(\frac{V_3}{V_1}\right) \tag{5.3}$$

   Here V_3 defines the rms value of 3rd harmonic in volts and V_1 defines the rms value of fundamental component of signal in volts.

   Total harmonic distortion (THD) is defined as a ratio of strength of all harmonic components to fundamental signal component. It is generally defined in dB as expressed in (5.4).

$$THD = \frac{\sqrt{V_2^2 + V_3^2 + V_4^2 + \ldots}}{V_1} \tag{5.4}$$

   Here V_n where n = {1, 2, 3} defines the rms value of nth harmonic voltage.

   Spurious free dynamic range (SFDR) is defined as a ratio of fundamental signal to largest spurious tone present in output. It is generally defined in dB.

The design of amplifier is tasked with achieving all of these specification where most of these specification involves trade-off and come at the expense of other. Hence this process is rather iterative where achieving an optimized design is often a challenging problem in hand.

5.3 IDENTIFICATION OF DEVICE PARAMETERS FOR AMPLIFIER DESIGN

Now if we are tasked with designing a voltage amplifier than one simple question arise is what kind of device can work as an amplifier. In this section we will attempt to answer this question. Ideally an amplifier is a linear two port network which can be modelled using any two port network parameter such as Y parameter.

Now consider an amplifier as a black box which is a two port network. Its input port is connected to a signal source which has a series conductance G_S. Its output is connected to a load having conductance G_L as shown in Fig. 5.1. In order to solve for its gain $\frac{v_o}{v_i}$, one can use KCL, KVL and device equations.

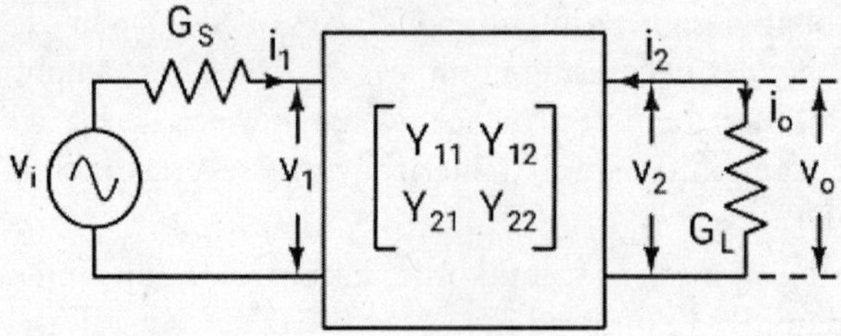

Fig. 5.1: Amplifier circuit with amplifier modelled as a two port network represented using Y-parameters

Writing KVL around leftmost loop in the circuit at input side we get the expression of v_1 given in (5.5)

$$v_1 = v_i - \frac{i_1}{G_S} \tag{5.5}$$

Similarly output voltage is defined in (5.6) using KVL,

$$v_o = v_2 = \frac{i_o}{G_L} = -\frac{i_2}{G_L} \tag{5.6}$$

Equations defining port voltages and currents are written using Y parameters as given in (5.7) and (5.8)

$$i_1 = Y_{11}\, v_1 + Y_{12}\, v_2 \tag{5.7}$$

$$i_2 = Y_{21}\, v_1 + Y_{22}\, v_2 \tag{5.8}$$

Using (5.5),(5.6) and (5.7) we get (5.9)

$$G_S\, v_i = (Y_{11} + G_S)v_1 + Y_{12}\, v_o \tag{5.9}$$

Further using (5.6) and (5.8) we get (5.10)

$$(Y_{22} + G_L)v_o = (Y_{21})v_1 \tag{5.10}$$

On eliminating v_1 using (5.9) and (5.10) we get (5.11)

$$A_v = \frac{v_o}{v_i} = \frac{Y_{21}G_S}{(Y_{22} + G_L)(Y_{11} + G_S) + Y_{12}Y_{21}} \tag{5.11}$$

Now in order to maximize the voltage gain A_v one need to maximize numerator or minimize denominator. The knobs available to us are Y parameters of amplifier. The numerator has only one Y parameter, Y_{21} and hence we can maximize it. Ideally one would like to choose $Y_{21} = \infty$ but in reality we can make it very large. However, Y_{21} is also a part of denominator which is not a good thing to maximize the gain. Luckily it is a part of product term where other parameter is Y_{12}. So choosing $Y_{12} = 0$ minimizes the denominator. In similar way choosing Y_{11} and Y_{22} as zero also minimizes the denominator which maximizes the voltage gain.

Thus ideal voltage amplifier should have following Y parameters given by (5.12)

$$Y = [00 \infty 0] \tag{5.12}$$

It results in following I-V equations given in (5.13) and (5.14)

$$i_1 = 0 \tag{5.13}$$

$$i_2 = Y_{21} v_1 \tag{5.14}$$

Thus a potential device which can perform amplification should have high input impedance. Further, its output current should be a function of its input voltage only. In real world having such a device which exactly meet all of these criteria might not be possible. However, one can aim to find a device with large Y_{12}, small Y_{11} and Y_{22} and almost negligible Y_{12}. It is very important that Y_{12} of device should be exceptionally small to achieve a large gain. Thus to a large extent such an amplifier is modelled using the model shown in Fig.5.2.

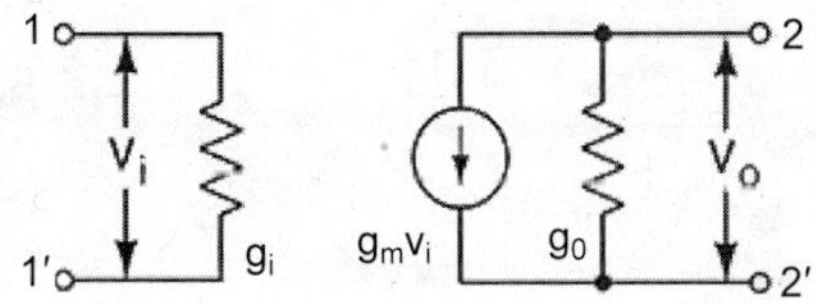

Fig. 5.2: Model of an amplifier

We can recall that small signal model of BJT and MOS transistor show this behaviour. Hence, BJT and MOS transistor are suitable choices for designing amplifiers. This analysis also makes it clear that these devices are designed in a way so as to achieve desired I-V characteristics and it is not by an accident that these devices are found to be useful for amplification.

5.4 FEEDBACK

When a part of amplifier output is fed back to it then output of amplifier becomes a function of its input as well output then such an amplifier is known as feedback amplifier or closed loop amplifier. Feedback is an important concept in designing amplifier hence we will devote some time in understanding it.

The feedback is the process of feeding system output back to the system. Feedback is an important aspect of system behaviour and may alter its properties. Hence it is important to analyze the effect of feedback in a system.

5.4.1 Open Loop System

An open loop system is a system which does not have any feedback in it. Fig.5.3(a) depicts an open loop system where system input is X(s), its output is Y(s) and its transfer function is G(s). It is assumed here that system described using G(s) does not have any feedback in it. Its transfer function $T_o(s)$ is defined using (5.15)

$$T_o(s) = \frac{Y(s)}{X(s)} = G(s) \tag{5.15}$$

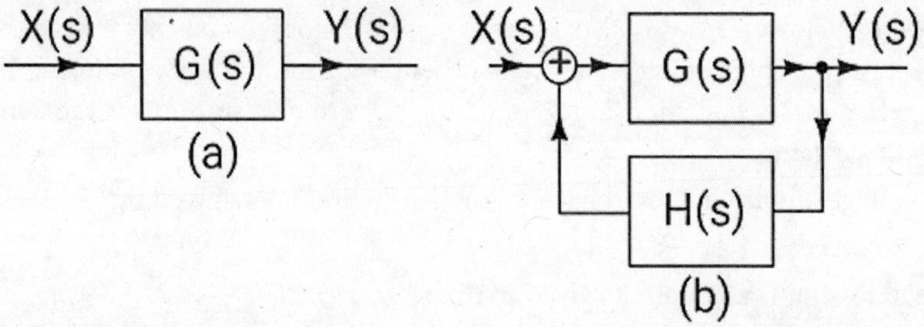

Fig. 5.3: (a) Open loop amplifier (b) Closed loop amplifier

Stability of a system

A system is unstable if any of its poles of transfer function are in right half of s plane (including imaginary axis). This is so because natural response of such system does not converge which makes system behaviour unstable. Hence, for a system to be stable all of its poles should be in left half s plane only.

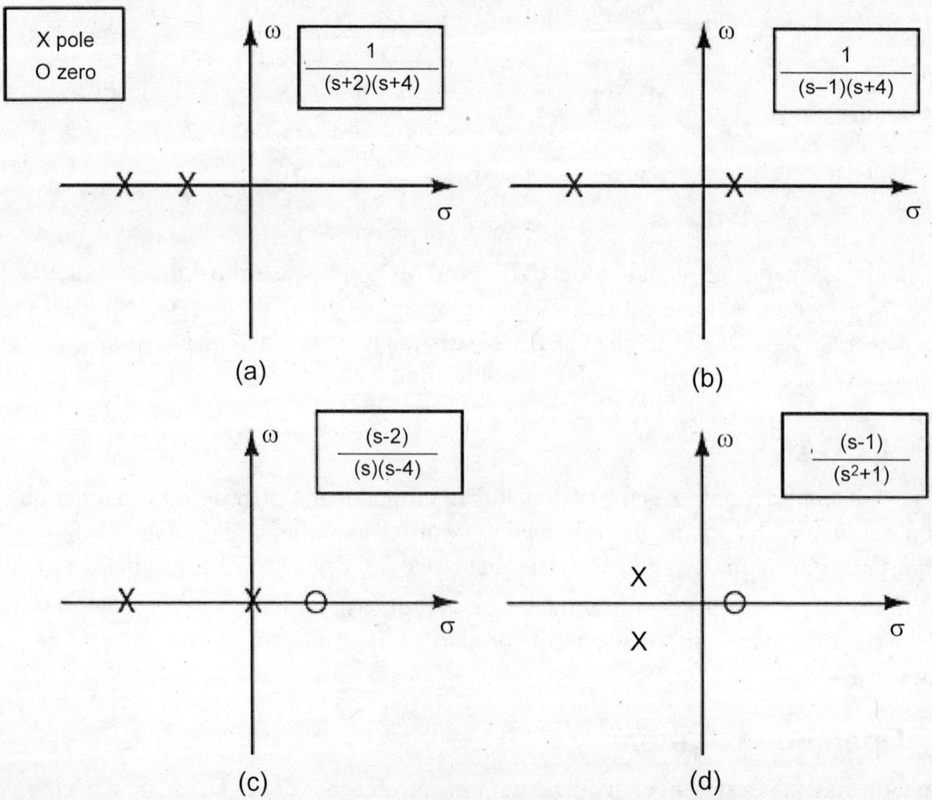

Fig. 5.4: (a) System having all poles in LHP (b) System with one pole in RHP (c) System with one pole on origin (d) System with complex poles in LHP

Fig.5.4 shows the position of poles and zeros of system in s-plane. The s-plane is a complex plane where s = σ + jω, here, s is a complex frequency, σ is real part of complex frequency and ω is imaginary part of complex frequency. The real part σ defines the exponential decay or growth of the natural response and imaginary part defines the oscillatory behaviour of natural response. The system shown in Fig.5.4(a) is a stable system as its all poles are in LHP. The system shown in Fig.5.4(b) is an unstable system as one of its poles are in RHP. The system shown in Fig.5.4(c) is an unstable system as one of its pole is on origin and not in LHP. The system shown in Fig.5.4(d) is a stable system as its both poles are in LHP. Although its poles are complex but as it is in LHP so the system is stable. One can notice that this system has its zero in RHP but that does not influence the stability of system. Hence this system is stable.

5.4.2 Closed Loop System

A system with feedback has at least one closed loop due to feedback. Hence, such systems are also known as closed loop systems. The Fig.5.3(b) depicts a closed loop system or a feedback system where G(s) is defined as its forward transfer function, H(s) is defined as its feedback transfer function. Its closed loop transfer function $T_c(s)$ is given using (5.16).

$$T_c(s) - \frac{Y(s)}{X(s)} = \frac{G(s)}{1-G(s)H(s)} \tag{5.16}$$

Here the term G(s)H(s) is defined as loop gain L(s) as it is the gain while travelling around the loop involving feedback. If loop gain is positive then such system are defined as positive feedback system while on the other hand if loop gain is negative then such feedback system are defined as negative feedback system.

In initial days, positive feedback were exploited to design amplifier with much higher gain than its open loop version. However, sensitivity of positive feedback is high and it is challenging to keep system remain stable across all possible variation. The negative feedback on the other hand reduces the gain but makes the system performance more robust with respect to variations.

5.5 CLASSIFICATION OF FEEDBACK AMPLIFIERS

In designing a feedback amplifier, feedback is taken from output port of the network and provided back to input port of the network.

The feedback signal can be extracted as a current or a voltage. Also it can be fed back to the original network either as a current or a voltage. As here two ports are involved and two possible ways of extracting or feeding back signal exists. Hence there are four possible configurations of designing such feedback amplifiers.

Following are these four configuration of feedback amplifier.

1. Current series feedback amplifier
2. Current shunt feedback amplifier
3. Voltage series feedback amplifier
4. Voltage shunt feedback amplifier

The naming convention of these configurations are as follows: the first term used in the name of configuration is the electrical quantity which is extracted from the original network to be used as a feedback signal in the feedback network. The second term describes how the output of the feedback network is connected to the input port of the original network.

A current feedback is extracted by connecting the input port of the feedback network in series with the output port of the original network because currents are same when two ports are connected in series. A voltage feedback is extracted by connecting the input port of the feedback network in shunt with the output port of the original network because voltages of two ports are same when it is connected in parallel. However, to provide the feedback back to the original network as a current the output port of feedback network is connected in parallel with the input port of the original network. This is done so because currents are added when connected in parallel. A voltage feedback is provided by connecting the output port of the feedback network in series with the input port of the original network.

The circuit model of amplifier shown in Fig.5.2 is used as an open loop amplifier to explain these feedback amplifiers. It has following specifications.

Its DC gain is given using (5.17)

$$Gain = -\frac{g_m}{g_o} \tag{5.17}$$

Its input impedance is given using (5.18)

$$Inputimpedance(R_{in}) = \frac{1}{g_i} \tag{5.18}$$

Its output impedance is given using (5.19)

$$Outputimpedance(R_{out}) = \frac{1}{g_o} \tag{5.19}$$

5.4.1 Current Series Feedback Amplifier

The Fig.5.5 shows the block diagram of a current shunt feedback amplifier. In this amplifier current from the output port of the original network is extracted by connecting the input port of the feedback network in series with the output port of the original network. The output voltage of the feedback network is fed back to the original network by connecting the output port of feedback network in series with the input port of the original network as shown in Fig.5.5.

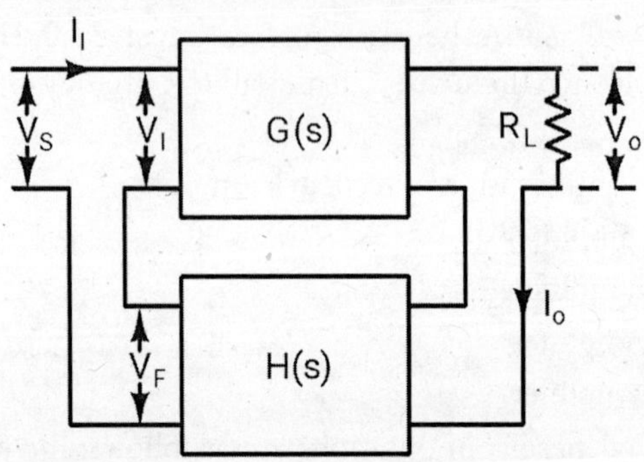

Fig. 5.5: Block diagram of current series feedback amplifier

The Fig.5.6 shows the equivalent circuit of current series feedback amplifier when model of open loop amplifier is used from Fig.5.2.

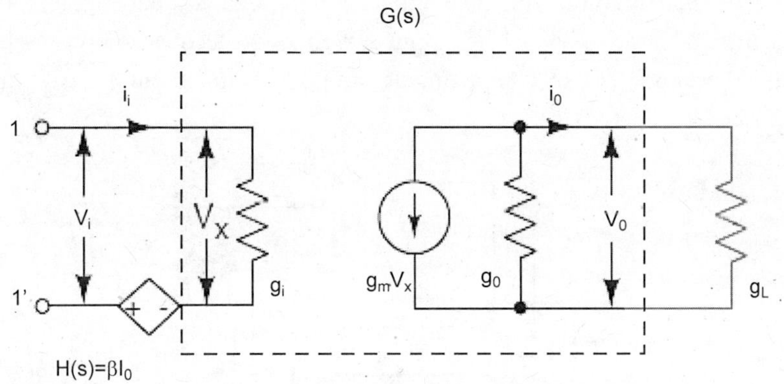

Fig. 5.6: Equivalent circuit of current series feedback amplifier on using amplifier model of Fig.5.2

On writing KCL around output node one can find the expression of v_o as shown in (5.20)

$$v_o = \frac{i_o}{g_L} = \frac{-g_m v_x}{g_o + g_L} \tag{5.20}$$

Similarly on writing KVL at input side one can find expression of v_x as shown in (5.21)

$$v_x = v_i + \beta i_o = v_i + \beta v_o g_L \tag{5.21}$$

Thus one can drive the gain of closed loop amplifier using (5.20) and (5.21) which is given in (5.22)

$$A_{cl} = \frac{v_o}{v_i} = \frac{-g_m / g_o}{1 + (g_L / g_o) + \beta(g_m / g_o) g_L} \tag{5.22}$$

The input impedance of current series amplifier is given using (5.23)

$$R_{in} = \frac{v_i}{i_i} = \frac{v_x - \beta i_o}{i_i} = \frac{v_x + v_x \beta \frac{g_m g_L}{g_o + g_L}}{i_i} = \frac{1 + \beta \frac{g_m g_L}{g_o + g_L}}{g_i} \tag{5.23}$$

Thus we can see that input impedance of current series amplifier is much larger than input impedance of original amplifier.

The output impedance of current series amplifier is given using when v_i is considered zero. In this case input v_x becomes equal to βi_o. Thus KCL at the output node can be written as shown in (5.24) to find the output impedance R_{out}.

$$i_o = -(g_m \beta i_o + v_o g_o) \Rightarrow R_{out} = -\frac{v_o}{i_o} = \frac{1 + g_m \beta}{g_o} \tag{5.24}$$

Thus output impedance of current series amplifier is higher than output impedance of original amplifier.

5.5.2 Current Shunt Feedback Amplifier

The Fig.5.7 shows the block diagram of a current shunt feedback amplifier. In this amplifier current from the output port is extracted by connecting the input port of the feedback network H(s) in series with the output port of the original network G(s). The output current of the feedback network H(s) is fed back to input port of original network G(s) by connecting output port of H(s) in parallel with input port of G(s) as shown in Fig.5.7.

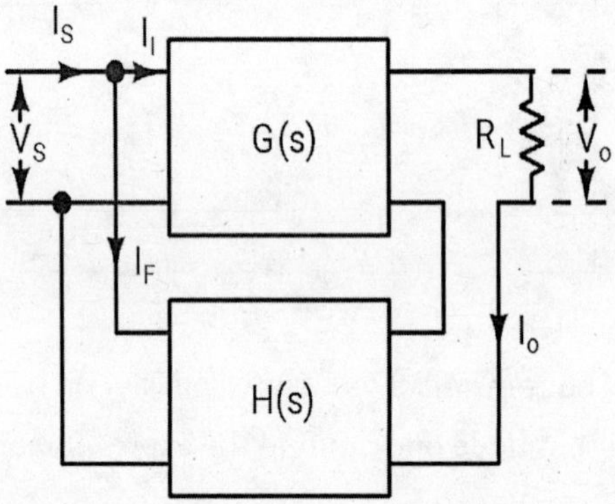

Fig. 5.7: Current shunt feedback amplifier

The Fig.5.8 shows the equivalent circuit of current shunt feedback amplifier when model of open loop amplifier is used from Fig.5.2.

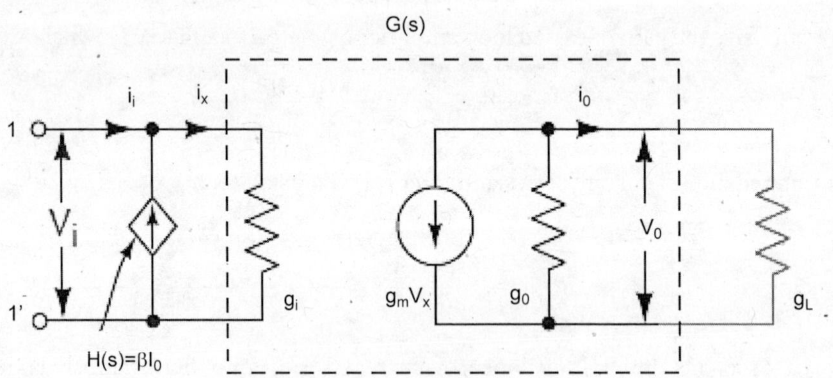

Fig. 5.8: Equivalent circuit of current shunt feedback amplifier on using amplifier model of Fig.5.2

The input impedance of current shunt amplifier is given in (5.25)

$$v_i = \frac{i_i + \beta i_o}{g_i} = \frac{i_i - \beta v_i \dfrac{g_m g_L}{g_o + g_L}}{g_i} \Rightarrow R_{in} = \frac{v_i}{i_i} = \frac{1}{g_i + \dfrac{\beta g_m g_L}{g_o + g_L}} \tag{5.25}$$

Thus if g_m is assumed to be sufficiently high and $g_o << g_I$ then it approaches to $1/\beta g_m$ which is low impedance.

Thus we can see that input impedance of current shunt amplifier is low impedance and hence this can be used as a current input amplifier.

The output impedance of current series amplifier is given using when v_i is considered zero. In this case input v_x becomes equal to βi_o. Thus KCL at the output node can be written as shown in (5.26) to find out output impedance R_{out}.

$$i_o = -(g_m\beta i_o + v_o g_o) \Rightarrow R_{out} = -\frac{v_o}{i_o} = \frac{1+g_m\beta}{g_o} \tag{5.26}$$

Thus output impedance of current series amplifier is also much higher than output impedance of original amplifier.

Now in order to calculate the gain of this amplifier, one can first write KCL around output node to find the expression of v_o as shown in (5.27)

$$v_o = \frac{i_o}{g_L} = \frac{-g_m v_i}{g_o + g_L} \tag{5.27}$$

Similarly on writing KCL at input side one can find expression of v_x as shown in (5.28)

$$v_i = \frac{i_x}{g_i} = \frac{i_i + \beta i_o}{g_i} = \frac{i_i + \beta v_o g_L}{g_i} \tag{5.28}$$

Thus one can drive the gain of closed loop amplifier using (5.27) and (5.28) which is given in (5.29)

$$A_{cl} = \frac{v_o}{i_i} = \frac{1}{g_i}\left(\frac{-g_m}{g_o + g_L + (\beta g_m g_L / g_i))}\right) \tag{5.29}$$

5.4.3 Voltage Series Feedback Amplifier

The Fig.5.9 shows the block diagram of a voltage series feedback amplifier. In this amplifier voltage from the output port V_o is extracted by connecting the input port of the feedback network H(s) in shunt with the output port of the original network G(s). The output voltage of the feedback network H(s) is fed back to input port of original network G(s) by connecting output port of H(s) in series with input port of G(s) as shown in Fig. 5.9.

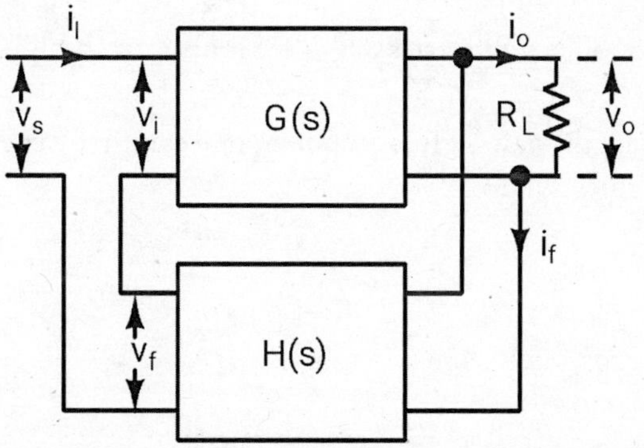

Fig. 5.9: Block diagram of voltage series feedback amplifier

The Fig.5.10 shows the equivalent circuit of voltage series feedback amplifier when model of open loop amplifier is used from Fig.5.2.

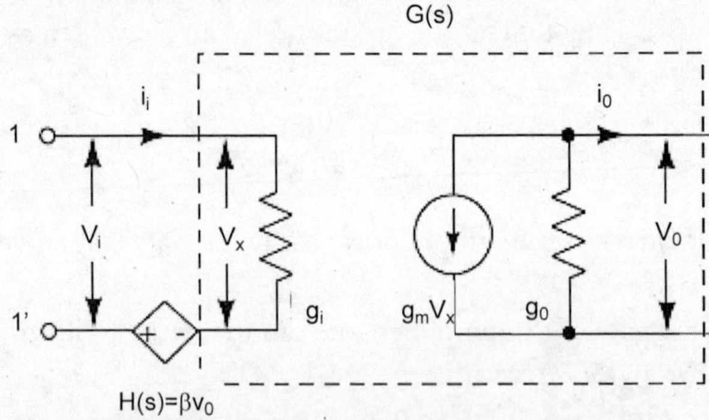

Fig. 5.10: Equivalent circuit of voltage series feedback amplifier on using amplifier model of Fig.5.2

The input impedance of voltage series amplifier is given in (5.30)

$$v_i = \frac{i_i}{g_i} - \beta v_o = \frac{i_i}{g_i} + \beta v_x \frac{g_m g_L}{g_o + g_L} = \frac{i_i}{g_i} + \beta \frac{i_i}{g_i} \frac{g_m g_L}{g_o + g_L} \tag{5.30}$$

Thus if g_m is assumed to be sufficiently high and $g_o << g_L$ then input impedance of voltage series feedback amplifier is much higher than original amplifier.

Thus we can see that input impedance of current shunt amplifier is high impedance and hence this can be used as a voltage input amplifier.

The output impedance of voltage series amplifier is calculated while considering v_i as zero. In this case input v_x becomes equal to βi_o. Thus KCL at the output node can be written as shown in (5.31) to find the output impedance R_{out}.

$$i_o = -(g_m \beta i_o + v_o g_o) => R_{out} = -\frac{v_o}{i_o} = \frac{1 + g_m \beta}{g_o} \tag{5.31}$$

Thus output impedance of current series amplifier is also much higher than output impedance of original amplifier.

Now in order to calculate the gain of this amplifier, one can first write KCL around output node to find the expression of v_o as shown in (5.32)

$$v_o = \frac{i_o}{g_L} = \frac{-g_m v_x}{g_o + g_L} = \frac{-g_m (v_i + \beta i_o)}{g_o + g_L} = \frac{-g_m (v_i + \beta g_L v_o)}{g_o + g_L} \tag{5.32}$$

It can be simplified to expression shown in (5.33)

$$A_{cl} = \frac{v_o}{v_i} = \frac{-g_m}{g_o + g_L + \beta g_m g_L} \tag{5.33}$$

5.5.4 Voltage Shunt Feedback Amplifier

The Fig.5.11 shows the block diagram of a voltage shunt feedback amplifier. In this amplifier voltage from the output port v_o is extracted by connecting the input port of the feedback network H(s) in shunt with the output port of the original network G(s). The output current of the feedback network H(s) is fed back to input port of original network G(s) by connecting output port of H(s) in shunt with input port of G(s) as shown in Fig.5.11.

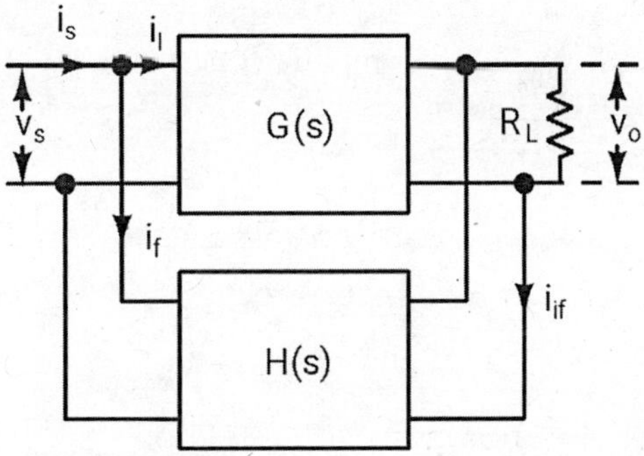

Fig. 5.11: Block diagram of voltage shunt feedback amplifier

The Fig.5.12 shows the equivalent circuit of voltage shunt feedback amplifier when model of open loop amplifier is used from Fig.5.2.

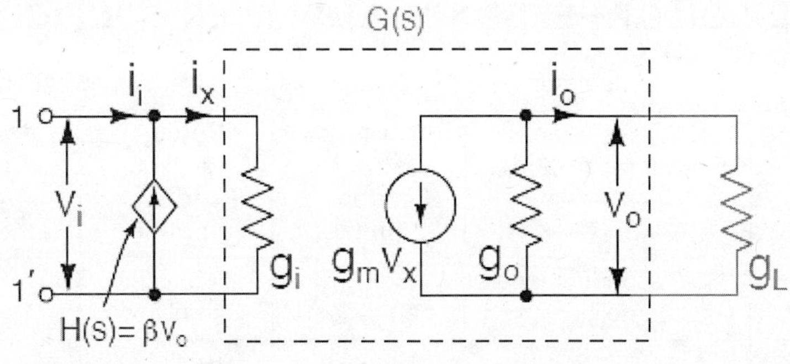

Fig. 5.12: Equivalent circuit of voltage shunt feedback amplifier on using amplifier model of Fig.5.2

The input impedance of voltage shunt amplifier is given in (5.34)

$$v_i = \frac{i_i + \beta v_o}{g_i} = \frac{i_i - \beta v_i \dfrac{g_m}{g_o + g_L}}{g_i} \Rightarrow R_{in} = \frac{v_i}{i_i} = \frac{1}{g_i + \dfrac{\beta g_m}{g_o + g_L}} \quad (5.34)$$

Thus if g_m is assumed to be sufficiently high then input impedance of voltage shunt feedback amplifier is much smaller than original amplifier.

Thus we can see that input impedance of voltage shunt amplifier is low impedance and hence this can be used as a current input amplifier.

The output impedance of voltage shunt amplifier is calculated while considering v_i as zero. In this case current flowing through g_i is βv_o. Thus KCL at the output node can be written as shown in (5.35) to find the output impedance R_{out}.

$$i_o = -(g_m \beta v_o + v_o g_o) \Rightarrow R_{out} = -\frac{v_o}{i_o} = \frac{1}{g_o + g_m \beta} \tag{5.35}$$

Thus output impedance of voltage shunt feedback amplifier is also much smaller than output impedance of original amplifier.

Now in order to calculate the gain of this amplifier, one can first write KCL around output node to find the expression of i_o as shown in (5.36)

$$v_o = \frac{i_o}{g_L} = \frac{-(g_m v_i + g_o v_o)}{g_L} = \frac{-(g_m \dfrac{i_i + \beta v_o}{g_i} + g_o v_o)}{g_L} \tag{5.36}$$

It can be simplified to expression shown in (5.37)

$$A_{cl} = \frac{v_o}{i_i} = \frac{-g_m}{g_i(g_o + g_L) + \beta g_m} \tag{5.37}$$

If g_m is assumed to be much larger than g_i, g_o and g_L, then closed loop gain approximately becomes -1/β

5.6 COMPARISON OF FEEDBACK AMPLIFIER CONFIGURATIONS

Table. 5.1: Comparison of different feedback amplifier configurations

Specifications	**Current series Feedback amplifier**	**Current shunt Feedback amplifier**	**Voltage series Feedback amplifier**	**Voltage shunt Feedback amplifier**
Nature of Amplifier	Transconductance amplifier	Current amplifier	Voltage amplifier	Transimpedance amplifier
Unit of gain	Mho	Unitless	Unitless	Ohm
Input impedance	Increase	Decrease	Increase	Decrease
Output impedance	Increase	Increase	Decrease	Decrease

5.7 STABILITY OF FEEDBACK OR CLOSED LOOP SYSTEM

If denominator in (5.16) become zero then magnitude of closed loop transfer function becomes infinity which allows system to have finite output even without having any input. This is the condition at which system becomes unstable and can oscillate. This condition L(s) = G(s)H(s) = 1 is also known as Barkhausen criterion. It is used to define condition of stability of a closed loop system. It is necessary but not a sufficient condition for oscillation.

According to Barkhausen criterion for a system to be oscillate its loop gain should exactly become unity. It implies that its magnitude response should become unity and its phase response should be a multiple of 2π at the frequency where gain becomes unity. However, in a practical world most of the systems are nonlinear and have compressive nonlinear behaviour which means gain can drop substantially when input is large. This means even if a system is stable for smaller inputs it can become unstable for larger inputs if phase becomes equal to or multiple of 2π.

For a negative feedback system, if L(s) = G(s)H(s) is defined without including the subtraction sign appearing in feedback loop which contributes180° or π rad phase, then its closed loop transfer function is given as expressed in (5.38)

$$T_c(s) = \frac{Y(s)}{X(s)} = \frac{G(s)}{1+L(s)} = \frac{G(s)}{1+G(s)H(s)} \tag{5.38}$$

The system defined using (5.3) can oscillates if phase contributed by L(s) becomes $(2k + 1)\pi$ where k is an integer. Thus to ensure the stability of a system we ensure that gain of system should become smaller than unity before it achieves a phase of π. We use a concept of phase margin to define the stability of a negative feedback system. The phase margin(PM) is defined as the change in phase of L(s) needed to make the system oscillate. If ϕ_u is the phase of the L(s) at unity gain frequency in radian/s, then phase margin(PM) of a negative feedback system is defined using (5.39)

$$\text{PM} = \pi - \phi_u \tag{5.39}$$

A positive phase margin defines a stable system and a negative phase margin defines an unstable system. Generally in most of the system phase margin is kept more than 30 degree or so to avoid strong ringing in its natural response.

In order to calculate the phase margin we can first find out the unity gain frequency of loop transfer function by equating |G(s)H(s)|(s = jω) = 1 and calculating the frequency ω_u from this. This frequency is defined as unity gain frequency (ω_u). Now phase margin is calculated using (5.40).

$$PM = \pi - \sum_{poles} \tan^{-1}\frac{Im(s-s_p)}{\text{Re}(s-s_p)}\Big|_{s=j\omega_u} + \sum_{zeros} \tan^{-1}\frac{Im(s-s_z)}{\text{Re}(s-s_z)}\Big|_{s=j\omega_u} \tag{5.40}$$

Here, *sp* denotes the frequency of the pole of loop gain and s_z denotes the frequency of the zero of the loop gain.

5.8 OSCILLATOR

Oscillator is a circuit which produces a periodic output signal. An oscillator does not require any input to produce an output signal. As we have earlier seen in (5.16) that a feedback system can have a finite output without having an input if its closed loop gain becomes infinity. This becomes possible when its loop gain L(s) becomes unity. Thus if a system satisfies this condition at any frequency other than DC then it can potentially oscillate at that frequency. The satisfaction of this condition at DC does not result in oscillation rather it results in an offset.

The condition of loop gain being unity translates into magnitude response of loop gain being unity and phase response being multiple of 2π radian. Thus as opposed to the design of amplifier where we want to avoid the case of loop gain being unity, this is in fact the desired case for achieving oscillation.

5.8.1 Phase Shift Oscillator

One common approach to design an oscillator is by designing a loop which has phase shift around the loop as multiple of 2π radian. Such kind of oscillator are also known as phase shift oscillators. There are several different possible methods of generating a phase shift. One common approach is to use an RC circuit to create a phase shift. A single RC circuit segment shown in Fig5.13(a) is a first order circuit which can provide a phase shift between 0° to 90° as shown in Fig.5.13(b). Several such RC circuit segments are cascaded to get a higher order transfer function which can provide a phase shift of more than 90°.

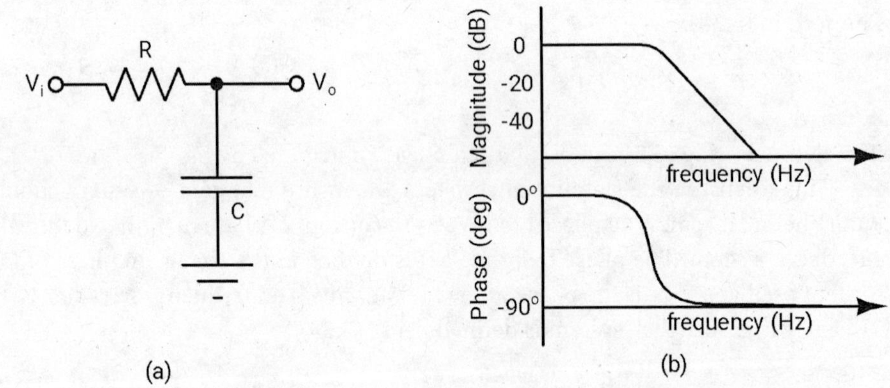

Fig. 5.13: (a) First order RC circuit (b) Magnitude and phase response of first order RC circuit shown in part(a)

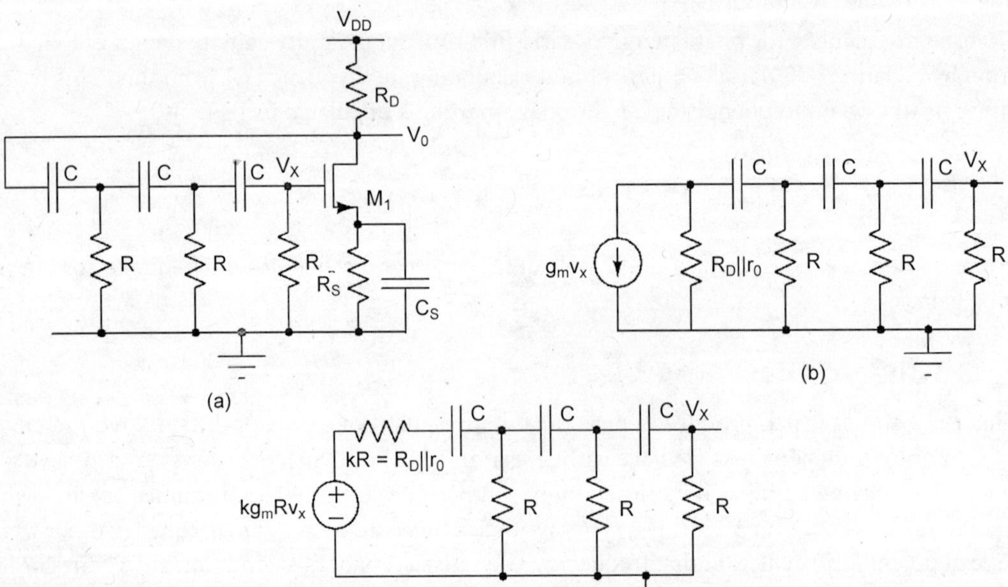

Fig. 5.14: (a) RC phase shift oscillator (b) Small signal equivalent of circuit shown in Fig.5.14(a). (c) Small signal equivalent of circuit shown in Fig.5.14(a) with Thevenin equivalent model of controlled source

As RC circuit segment is a passive circuit and hence it results in attenuation of the signal. Hence an amplifier is also needed in the loop to ensure the loop gain to be unity. The Fig.5.14 (a) shows an example of a phase shift oscillator circuit. Here transistor M_1 is used as an amplifier which also provides a 180° phase shift to the signal. The remaining180° phase shift is provided by the RC network connected between drain of M_1 and gate of M_1. In order to analyze this circuit, one can first draw the small signal equivalent of this circuit which is shown in Fig.5.14(b). The Norton equivalent of controlled current source is drawn as Thevenin equivalent in Fig.5.14(c) which also represent the small signal equivalent of circuit shown in Fig.5.14(a).

The loop gain of this circuit is given using (5.41)

$$L(s)=\frac{s^3kRg_mR^3C^3}{1+(5+k)sCR+(4k+6)s^2R^2C^2+(3k+1)s^3R^3C^3} \tag{5.41}$$

In order to calculate its frequency, first we substitute s with as shown in (5.42)

$$L(j\omega)=\frac{j\omega^3kRg_mR^3C^3}{1+j(5+k)\omega CR-(4k+6)\omega^2R^2C^2-j(3k+1)\omega^3R^3C^3} \tag{5.42}$$

(5.42) can be further simplified as (5.43)

$$L(j\omega)=\frac{\omega^3kRg_mR^3C^3}{-j+(5+k)\omega CR+j(4k+6)\omega^2R^2C^2-(3k+1)\omega^3R^3C^3} \tag{5.43}$$

Now for phase of L(jω) to be zero, imaginary part of its denominator should become zero as phase of numerator is already zero degree. This condition is given in (5.44)

$$-1+(4k+6)\omega^2R^2C^2=0 \Rightarrow \omega=\frac{1}{\sqrt{4k+6}\cdot RC} \tag{5.44}$$

In order to find the desired transconductance one can substitute this value of ω in (5.44) to get the expression shown in (5.45)

$$Rg_m=4k+\frac{29}{k}+23 \tag{5.45}$$

The oscillator acts as a backbone of majority of electronic system. It generates a timing signal use as a clock in digital circuits such as microprocessors or a reference signal to be used as a frequency reference in radio transmission and reception. The stability of this timing or frequency source is an important parameter otherwise any change in frequency of the source affects the performance of circuit such as transmission or reception in case of radio system. It can also interfere with the transmission or reception of the nearby radio channels if frequency deviates. This has forced the designers to look for stable frequency sources. The oscillators designed using crystals (mostly of quartz) which are also known as crystal oscillator are found to be highly stable frequency source. Hence, it is common to use crystal oscillator as a frequency reference in many applications.

5.8.2 Crystal Oscillator

A crystal is essentially a material where atoms or molecules are arranged in an orderly and repetitive pattern extending in all of its dimensions. Fig.5.15(a) shows the symbol of a crystal device. A crystal of quartz can be deformed or distorted when an electric field is applied across it. When the electric field is removed the material gets back to its original position and generates an electric potential across it. Now if these electrical signals are taken, amplified and fed back to the crystal then sustained oscillation can be built. The electrical behaviour of crystal resembles with RLC circuit and thus it is modelled using a RLC circuit. Its equivalent model is shown in Fig.5.15(b). Here the term C_s and L defines the mechanical characteristic of crystal, term R is the series resistance of circuit and Cp defines the capacitance between electrodes used to connect crystal. It shows series as well as parallel resonance behaviours. The frequency of series resonance is generally lower than parallel resonance frequency. The C_P is generally larger than C and also gets affected by other parasitic capacitance coming in parallel to circuit. Hence, oscillator designed using parallel resonance of crystal is less stable than the oscillator circuit designed using series resonance of the crystal. A crystal behaves like an inductive impedance between its series and parallel resonance. The quality factor of crystal is very high which makes it attractive for frequency reference application as it offers a high purity of frequency signal.

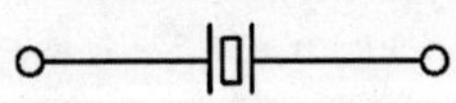

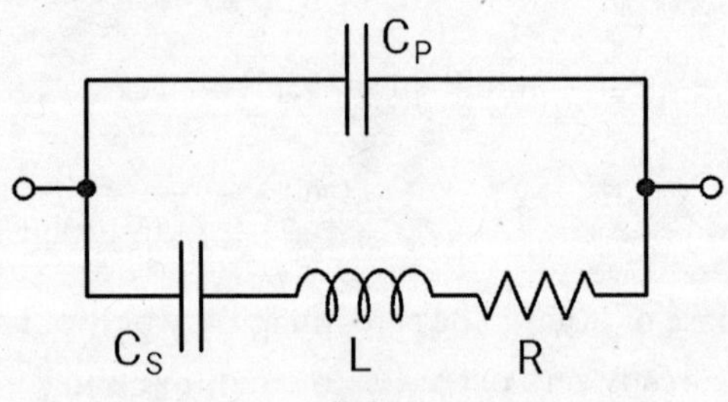

Fig. 5.15: (a) Symbol of a crystal (b) Electrical equivalent circuit of a crystal

The series resonance frequency f_s and parallel resonance frequency f_p of crystal is given using (5.46)

$$f_s = \frac{1}{2\pi\sqrt{LC_s}}, f_p = \frac{1}{2\pi\sqrt{L\frac{C_sC_p}{C_s+C_p}}} \tag{5.46}$$

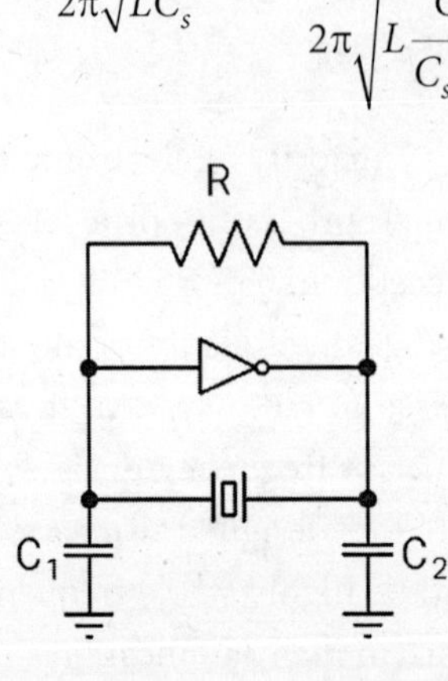

Fig. 5.16: Crystal oscillator in pierce oscillator configuration realized using CMOS inverter

Fig.5.16 shows the realization of crystal oscillator using pierce oscillator configuration. It uses parallel resonance mode of oscillation where series combination of C_1 and C_2 effectively constitute the C_{P1}. The value of these capacitors are chosen in such a way so that stability of this oscillator is preserved. For this C_{P1} should be larger than other expected parasitic capacitors. The resistive feedback is used to self-bias the inverter in its amplification region so that it can compensate for the losses of the RLC network.

5.8.3 Nonlinear Oscillator

The class of oscillator circuits which does not produce a sinusoidal output are known as nonlinear oscillator as it does not operate in linear or small signal region of operation. Examples of such oscillators are relaxation oscillator. The Fig.5.17(a) shows a circuit of hysteresis based relaxation oscillator. Here positive feedback creates a hysteresis while negative feedback generates a trigger to automatically switch the output from one state to another state of the oscillator. Thus this circuit is a astable multivibrator or an oscillator which produces a square waveform at its output as shown in Fig.5.17(b).

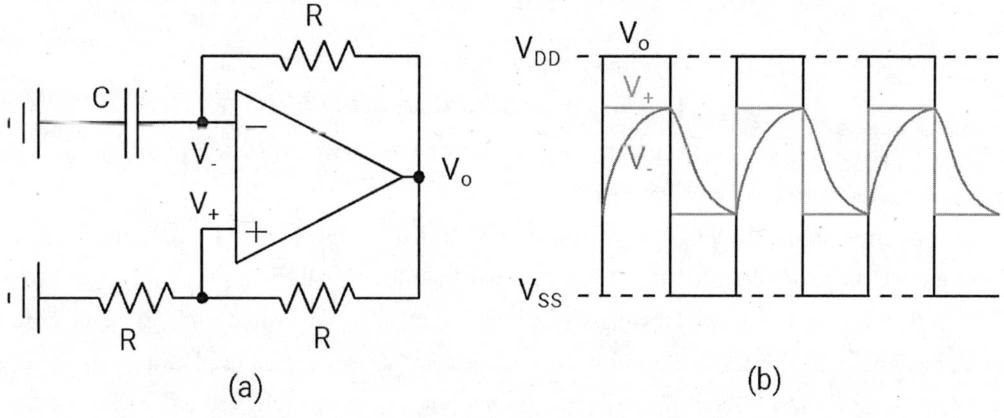

Fig. 5.17 (a) Circuit diagram of a hysteresis based relaxation oscillator (b) Waveforms at different nodes of circuit in Fig.5.17(a)

The time period of this oscillator is given using the (5.47)

$$T = RC\left(\ln\left(\frac{2V_{SS} - V_{DD}}{V_{SS}}\right) + \ln\left(\frac{2V_{DD} - V_{SS}}{V_{DD}}\right)\right) \quad (5.47)$$

Where V_{DD} is positive reference supply and V_{SS} is negative reference supply of the opamp.

Thus if $V_{DD} = V_{SS}$ then frequency of this oscillator is given by (5.48).

$$f = \frac{1}{2 \cdot RC \cdot \ln(3)} \quad (5.48)$$

5.8.4 Pulse oscillator

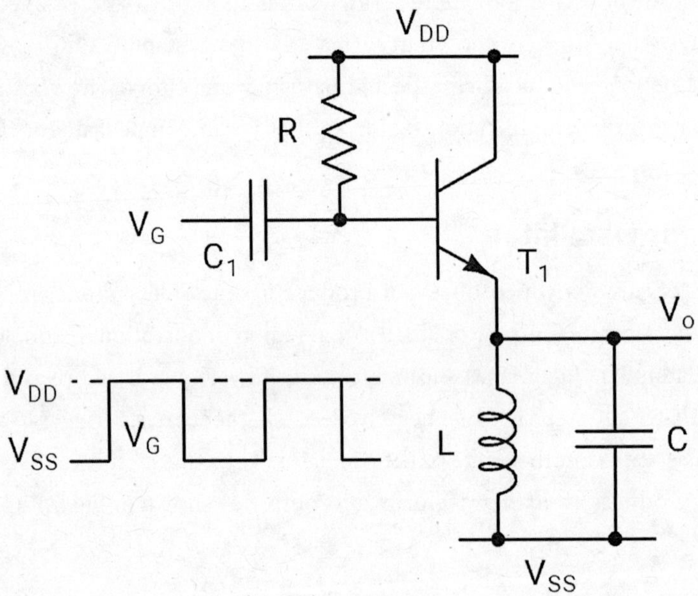

Fig. 5.18: Pulsed oscillator circuit

Certain applications such as radar need an oscillator circuit where the output of oscillator is not needed continuously rather in an intermittent way. Hence the oscillator circuit is designed to generate pulse of oscillating output. The Fig.5.18 shows the circuit of a pulsed oscillator circuit. Here an LC tank is connected to emitter of transistor T_1. This circuit works as follows. If the input at the gate of transistor T_1 is high then a large current flows through it and it kills any possible oscillation in LC tank. Now when gate of T_1 becomes low, T1 turns off and LC oscillator start generating an oscillating output. If quality factor of LC circuit is high enough then this oscillating output can last long enough or till the time gate input of T_1 become high again.

5.8.5 Ring Oscillator

Ring oscillator is a type of linear oscillator which is commonly used in on-chip realization of oscillator because of its easy implementation. A ring oscillator is essentially a ring of inverters where several inverters are cascaded by connecting output of one inverter to the input of next inverter and so on. The output of last inverter in this chain is finally connected to the input of the first inverter in the chain to complete the ring. Let's consider a simple case of ring oscillator where ring is only formed by one inverter. This case is shown in Fig.5.19(a) where output of this inverter is connected back to its input. The question asked in this case is what is the output waveform of voltage v_o? Some thinks that if initially input is logic zero or 0 then CMOS inverter should results in output as logic one or v_{dd}. Now this new logic one or v_{dd} at output goes back to the input as these terminals are shorted. The logic one or v_{dd}

at input forces the output to move to logic zero or 0. Thus output start oscillating between logic zero and logic one. However, this understanding is not correct. The flaw in this explanation is that we are assuming input and output are different at a particular time which is not possible as it is shorted. The small signal equivalent of this circuit shown in Fig.5.19(c) is a first order system which can provide a phase shift of 0 to 90 degree. The negative feedback from input to output provides an additional phase shift of 180 degree. Thus total phase shift possible in the loop is between 180 degree to 270 degree and thus this system cannot oscillate. The output voltage of this circuit can also be obtained by finding the intersection of the inverter transfer characteristic with line $v_o = v_i$ defining the short between input and output of inverter as shown in Fig.5.19(d).

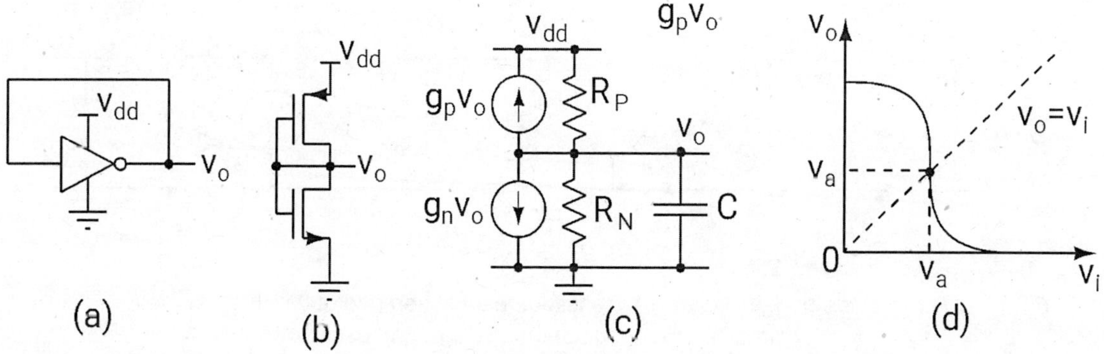

Fig.5.19: (a) CMOS inverter with feedback (b) Transistor level circuit of Fig.5.19(a) (c) Equivalent circuit of Fig.5.19(b) (d) Finding output voltage of Fig.5.19(c) using inverter characteristic and constraining it with input and output short ($v_o = v_i$) condition

If we take now use two inverters instead of one and connect the output of first inverter to the input of second inverter and then feed the output of second inverter back to the input of first inverter. This forms a ring of inverters. This circuit has two poles which provides a phase shift between 0 to 180 degrees at any frequency. However in this case the total phase shift contributed due to sign change is 360 degrees. Thus total phase shift in loop is between 360 degree to 540 degrees. Hence this system has a tendency to be unstable at DC because it can have an infinite gain at DC. However, having an infinite gain at DC may result in finite output at DC without having any input. The finite output at DC is essentially an offset and not an oscillation.

Now increase the size of ring by adding one more inverter as shown in Fig.5.20(a). The ring thus becomes a third order system. The detailed transistor level schematic of this ring is shown in Fig.5.20(b). The phase shift contributed by this third order system can be anywhere between 0 to 270 degree. The sign change due to inverter operation further contributes a phase shift of 540 degrees. Thus total phase shift in the ring is between 540 degree to 810 degree. Hence, phase shift can exactly be multiple of 2π which is 720 degree at some frequency and circuit can potentially oscillate at that frequency.

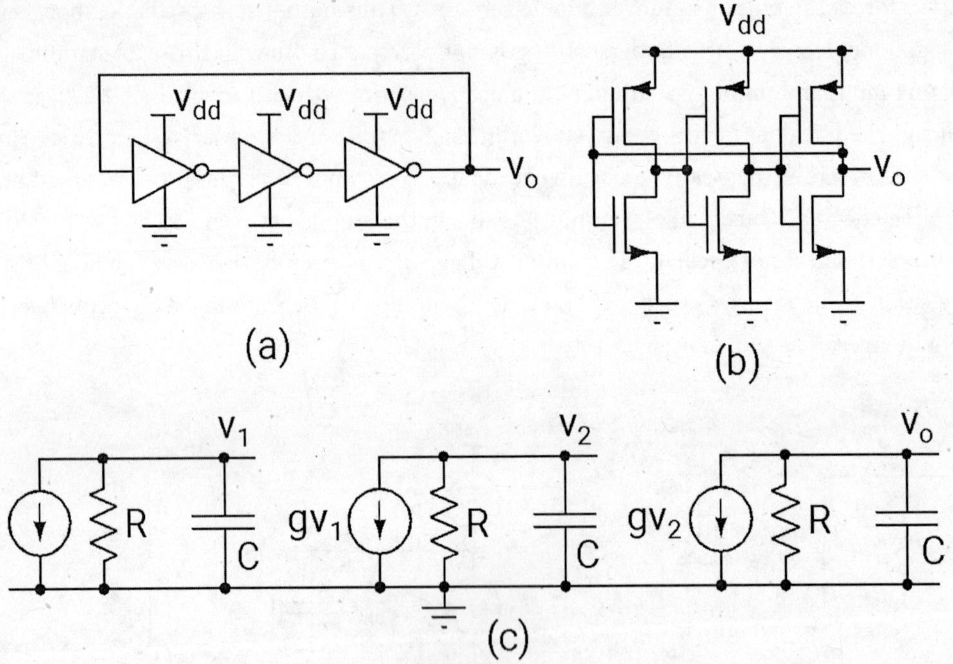

Fig. 5.20: (a) Three stage ring oscillator (b) Transistor level schematic of the three stage ring oscillator (c) Small signal circuit of three stage ring oscillator

The small signal equivalent of this circuit is shown in Fig.5.20(c). Its loop gain is given using (5.49)

$$H(s) = \frac{-(gR)^3}{(1+sRC)^3} \tag{5.49}$$

As transfer function already has a negative sign, hence contribution from all three poles if becomes then it can satisfy the condition of oscillation. The oscillation frequency of this circuit is given using (5.50)

$$\phi = 3\tan^{-1} \omega_{osc}RC = \pi \;=>\; \omega_{osc} = \frac{\tan\left(\frac{\pi}{3}\right)}{RC} = \frac{\sqrt{3}}{RC} \tag{5.50}$$

UNIT SUMMARY

This chapter has provided an introduction to amplifier and its design specification. It has also provided the basic definition of different design specification. The derivation of the two port parameters of an amplifier is done to identify the requirements on device for designing an amplifier. The chapter has introduced the concept of feedback and provided the details about different types of feedback such as positive and negative feedback. Different feedback amplifier configurations are discussed in detail. It has provided an introduction to oscillators and different types of oscillators providing an insight into its design details.

EXERCISES

Multiple Choice Questions

1. Which of the following amplifier specification defines second order nonlinearity?
(a) HD_2
(b) HD_3
(c) THD
(d) SFDR

2. Which of the following amplifier specification is not a nonlinearity specification?
(a) HD_2
(b) SFDR
(c) THD
(d) UGB

3. Which of the parameter should not be zero in an amplifier?
(a) Y_{11}
(b) Y_{12}
(c) Y_{21}
(d) Y_{22}

4. What should be the total phase shift in a loop to generate oscillation ?
(a) 2π
(b) π
(c) $\frac{\pi}{2}$
(d) $\frac{-\pi}{2}$

5. Which of the following device acts as an amplifier?
(a) Diode
(b) BJT
(c) Capacitor
(d) Inductor

6. Which of the following statement is true for negative feedback system?
(a) It provides a higher gain than the gain of open loop amplifier
(b) It results in higher distortion
(c) It always leads to instability
(d) It provides smaller gain than the gain of open loop amplifier

7. Which of the following oscillator is nonlinear?
(a) Ring oscillator
(b) crystal oscillator
(c) LC oscillator
(d) hysteresis relaxation oscillator

8. Which of the following statement is true?
(a) Negative feedback reduces the gain of the system
(b) Negative feedback increases the gain of the system
(c) Negative feedback increases the bandwidth of the system
(d) Negative feedback increases the distortion of the system

ANSWERS															
1.	(a)	2.	(d)	3.	(c)	4.	(a)	5.	(b)	6.	(d)	7.	(d)	8.	(a)

Short Answer Type Questions

1. Define signal to noise ratio?
2. What is Spurious free dynamic range?

3. What is feedback? What are the different types of feedback?
4. Write the statement of Barkhausen criterion of stability?
5. What are different feedback amplifier configuration?
6. What is oscillator? What are different types of oscillator?
7. What is ring oscillator?

Long Answer Type Questions

1. Discuss in detail about the different specifications of the amplifier?
2. Compare the different types of feedback amplifier configuration.
3. Define the condition of oscillation of a circuit.

Numerical Problems

1. SNR of input signal of an amplifier is 50 dB. Assume amplifier is noiseless and its gain is 40dB. What will be the SNR at its output?
2. An amplifier has its HD2=-70dB, HD3=-80dB and HD5=-84dB and all other harmonics are less than -100dB then what is its approximate THD?

PRACTICAL

Experiment-5.1: Simulation of Voltage Series and Voltage Shunt Feedback Amplifiers

Aim: Following are the aim of this simulation

1. Find operating points and small signal parameters of the transistor
2. Plot the AC magnitude and phase response of the amplifier with and without feedback
3. Plot transient response of the amplifier with and without feedback
4. Plot step response of the amplifier with and without feedback
5. Find input impedance, output impedance and gain of the designed amplifier

Equipment and Components:

S.No.	Name of the component	Specification	Quantity
1	Computer		
2	Spice Simulator	LTspice or any other spice simulator	
3	PDK models	Any technology node such as 180nm	1

Theory related to simulation :

Simulation is a method to analyze the behaviour or performance of a circuit using a computer without realizing it on hardware. The circuit is described to a computer using languages such as SPICE which is an acronym for Simulation Program with an Integrated Circuit Emphasis. A sample SPICE program looks like following,

Title of the circuit
Description of the circuit (This part may extend up to several lines)
.MODEL STATEMENTS
ANALYSIS COMMANDS
OUTPUT COMMANDS
.END

Here first line of the program defines the title of the circuit. This line is not considered during compilation and printed directly to the display output. After this next few lines describes the circuit. This part of program is termed as netlist. Thereafter, the path of model files are provided using .MODEL keyword. The model files provide information about behaviour of device. This modelling can either be done in hardware description language for behavioural modelling or in SPICE for experimental models. These model files are included in netlist with .MODEL keyword to define the operation of device. The analysis commands are used to define different simulation methods to analyze the performance of circuit. For example operating point analysis is performed to obtain the information about the operating points of the circuit. The DC analysis is performed to analyze the variation in operating points if one of the circuit parameter varies. The transient analysis is performed to obtain the time domain behaviour of the circuit where a time domain input is applied to the circuit and its behaviour is obtained as a function of time. The AC analysis is a small signal analysis where the circuit is first linearized around the operating point and then this linear version of circuit is simulated across different frequency to obtain the frequency response of the circuit. The AC analysis is always linear so magnitude of applied input signal is not much important if one want to predict the gain of the system using this analysis. The noise analysis is also a small signal analysis where small signal version of circuit is analyzed with noise models of devices. There are many other simulation methods which are not much important for the experiment discussed here but its information can be easily obtained through the resources available online. Thereafter set of output commands are used to display results obtained from simulations.

Procedure:

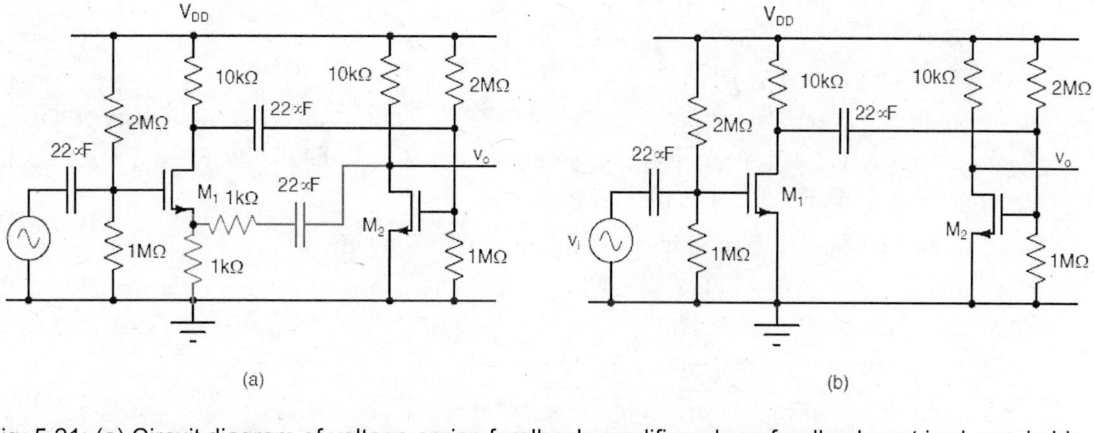

Fig. 5.21: (a) Circuit diagram of voltage series feedback amplifier where feedback part is shown in blue (b) Circuit diagram of amplifier circuit shown in part(a) with feedback element removed

1. Download a circuit simulator such as LT-Spice which can be downloaded at no cost from internet.
2. Launch the circuit simulator and create a new schematic file. Open this newly created schematic file.
3. Now the objective is to create a schematic of the circuit shown in Fig.5.21(a). This is done by first instantiating all the components used in this circuit. The components are instantiated by using Add component option available in LT spice.
4. Now connect the components to create the circuit shown in Fig.5.21(a).
5. The input source parameters are specified based on desired simulation. For operating point analysis select it as DC type and specify its DC value. For transient analysis, select if as Sine from styles and specify its DC value, amplitude and frequency. For AC analysis its small signal parameters are specified. As small signal analysis is linear so selecting AC magnitude as 1 is convenient as in this case magnitude response directly provides the gain of the amplifier.
6. The behaviour of the components are described using model file. The model file is a part of process development kit (PDK) which can be downloaded from website of respective foundry or supplier of the component. The model file is added by inserting a spice directive .INCLUDE <path> in the schematic window
7. Now again use SPICE directives to add the analysis commands to schematic window.

   The command to add operating point analysis is following.

   .op

   The command to add transient analysis is following

   .TRAN STEP STOP START

   Here STEP is the time step used by simulator to perform analysis and should be smaller than minimum expected event time. The STOP is the time up to which simulation is performed. The START is the time at which simulation begins.

   The command to add AC analysis is following

   .ac DEC NPOINTS FSTART FSTOP

   Here NPOINTS is number of points and DEC defines decade. Thus DEC along with NPOINTS defines that every decade of frequency has number of points equal to NPOINTS. FSTART is the starting frequency and FSTOP is the end frequency.
8. Now simulate the circuit by clicking over the run simulation button.
9. Simulation results are available in *.raw file which can be opened using waveform window. The simulated data can then be seen by selecting appropriate data and plotting it. The small signal information is available in log files and not directly in waveform window.
10. Plot the results and save the waveforms.
11. Repeat the steps for the amplifier shown in Fig.5.21(b) which provides the parameters of amplifier for the case of without feedback.

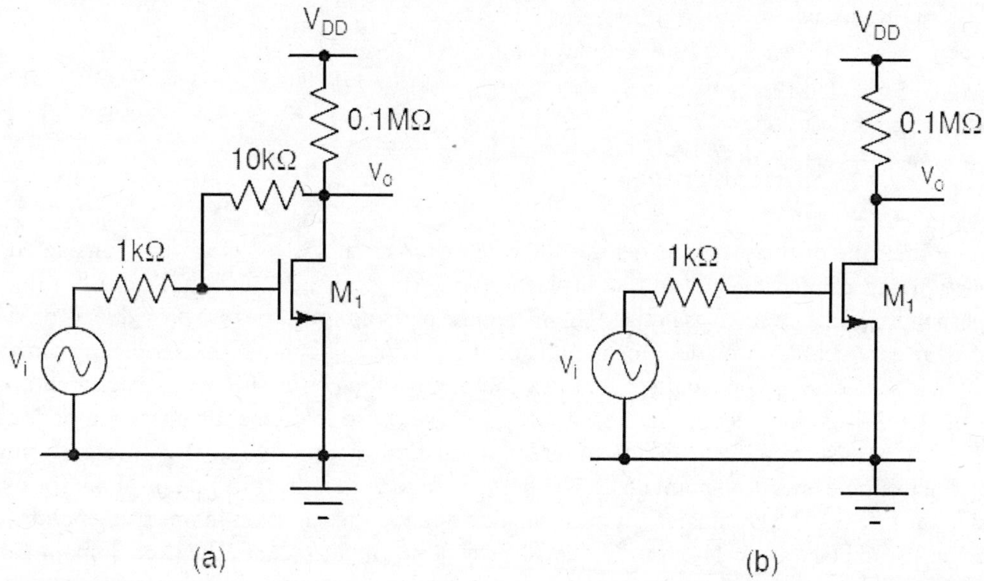

Fig. 5.22: (a) Circuit diagram of voltage shunt feedback amplifier (b) Circuit diagram of voltage series feedback amplifier with feedback element removed

12. Repeat the steps from S. No. 3 to S. No. 11 for the circuit shown in Fig.5.22.

Experiment-5.2: Simulation of Current Series and Current Shunt Feedback Amplifiers

Aim: Following are the aim of this simulation

1. Find operating points and small signal parameters of the transistor
2. Plot the AC magnitude and phase response of the amplifier with and without feedback
3. Plot transient response of the amplifier with and without feedback
4. Plot step response of the amplifier with and without feedback
5. Find input impedance, output impedance and gain of the designed amplifier

Equipment and Components:

S.No.	Name of the component	Specification	Quantity
1	Computer		
2	Spice Simulator	LTspice or any other spice simulator	
3	PDK models	Any technology node such as 180nm	1

Theory Related to Simulation:

Simulation is a method to analyze the behaviour or performance of a circuit using a computer without realizing it on hardware. The circuit is described to a computer using languages such as SPICE which is an acronym for Simulation Program with an Integrated Circuit Emphasis. A sample SPICE program looks like following,

Title of the circuit

Description of the circuit (This part may extend up to several lines)

.MODEL STATEMENTS

ANALYSIS COMMANDS

OUTPUT COMMANDS

.END

Here first line of the program defines the title of the circuit. This line is not considered during compilation and printed directly to the display output. After this next few lines describes the circuit. This part of program is termed as netlist. Thereafter, the path of model files are provided using .MODEL keyword. The model files provide information about behaviour of device. This modelling can either be done in hardware description language for behavioural modelling or in SPICE for experimental models. These model files are included in netlist with .MODEL keyword to define the operation of device. The analysis commands are used to define different simulation methods to analyze the performance of circuit. For example operating point analysis is performed to obtain the information about the operating points of the circuit. The DC analysis is performed to analyze the variation in operating points if one of the circuit parameter varies. The transient analysis is performed to obtain the time domain behaviour of the circuit where a time domain input is applied to the circuit and its behaviour is obtained as a function of time. The AC analysis is a small signal analysis where the circuit is first linearized around the operating point and then this linear version of circuit is simulated across different frequency to obtain the frequency response of the circuit. The AC analysis is always linear so magnitude of applied input signal is not much important if one want to predict the gain of the system using this analysis. The noise analysis is also a small signal analysis where small signal version of circuit is analyzed with noise models of devices. There are many other simulation methods which are not much important for the experiment discussed here but its information can be easily obtained through the resources available online. Thereafter set of output commands are used to display results obtained from simulations.

Procedure:

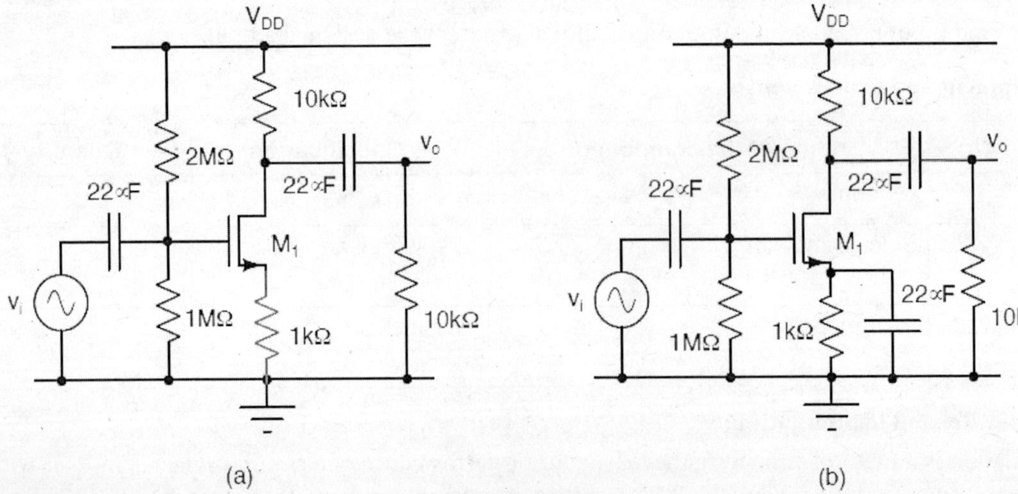

Fig. 5.23: (a) Circuit diagram of current series feedback amplifier (b) Circuit diagram of current series feedback amplifier with feedback element bypassed

1. Download a circuit simulator such as LT-Spice which can be downloaded at no cost from internet.
2. Launch the circuit simulator and create a new schematic file. Open this newly created schematic file.
3. Now the objective is to create a schematic of the circuit shown in Fig.5.23(a). This is done by first instantiating all the components used in this circuit. The components are instantiated by using Add component option available in LT spice.
4. Now connect the components to create the circuit shown in Fig.5.23(a).
5. The input source parameters are specified based on desired simulation.
6. For operating point analysis select it as DC type and specify its DC value. For transient analysis, select if as Sine from styles and specify its DC value, amplitude and frequency. For AC analysis its small signal parameters are specified. As small signal analysis is linear so selecting AC magnitude as 1 is convenient as in this case magnitude response directly provides the gain of the amplifier.
7. The behaviour of the components are described using model file. The model file is a part of process development kit (PDK) which can be downloaded from website of respective foundry or supplier of the component. The model file is added by inserting a spice directive .INCLUDE <path> in the schematic window
8. Now again use SPICE directives to add the analysis commands to schematic window.
9. The command to add operating point analysis is following.

   .op

   The command to add transient analysis is following

   .TRAN STEP STOP START

   Here STEP is the time step used by simulator to perform analysis and should be smaller than minimum expected event time. The STOP is the time up to which simulation is performed. The START is the time at which simulation begins.

   The command to add AC analysis is following

   .ac DEC NPOINTS FSTART FSTOP

   Here NPOINTS is number of points and DEC defines decade. Thus DEC along with NPOINTS defines that every decade of frequency has number of points equal to NPOINTS. FSTART is the starting frequency and FSTOP is the end frequency.

9. Now simulate the circuit by clicking over the run simulation button.
10. Simulation results are available in *.raw file which can be opened using waveform window. The simulated data can then be seen by selecting appropriate data and plotting it. The small signal information is available in log files and not directly in waveform window.
11. Plot the results and save the waveforms.
12. Repeat the steps for amplifier shown in Fig.5.21(b) which provides the simulation results for the case of without feedback.

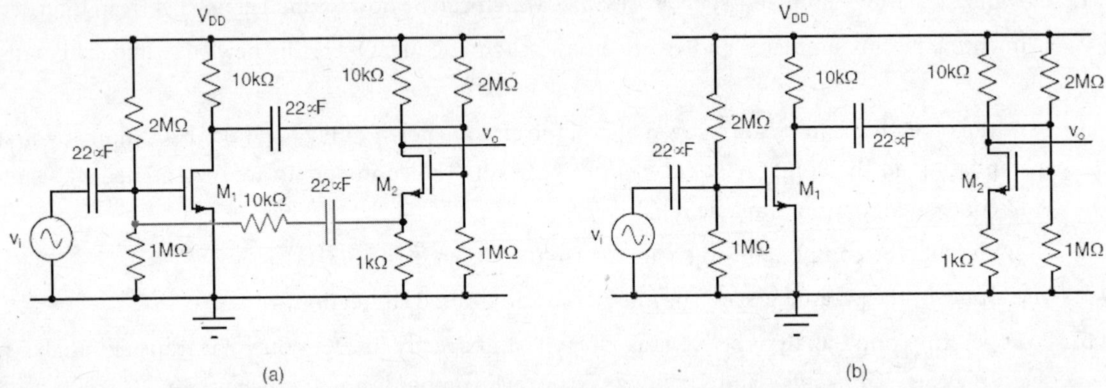

Fig. 5.24: (a) Circuit diagram of current shunt feedback amplifier (b) Circuit diagram of amplifier in part(a) with feedback element removed

13. Repeat the steps from S. No. 3 to S. No. 11 for the circuit shown in Fig.5.24.

Experiment-5.3: Study of Voltage Series and Voltage Shunt Feedback Amplifiers and plot the output. Compare the results with simulation model.

Aim: Following are the aim of this simulation

1. Find the gain of amplifier with and without feedback.
2. Find the AC magnitude and phase response of the amplifier with and without feedback.
3. Plot transient response of the amplifier with and without feedback
4. Plot step response of the amplifier with and without feedback.
5. Compare the results with simulation results of experiment 5.1.

Equipment and Components:

S.No.	Name of the component	Specification	Quantity
1	NPN transistor	BC547	2
2	Resistors	1kΩ	2
3	Resistors	10kΩ	2
4	Resistors	2MΩ	2
5	Resistors	1MΩ	2
6	Capacitors	22µF	3
7	Breadboard		1
8	Signal source		1
9	Power supply		1

Procedure:

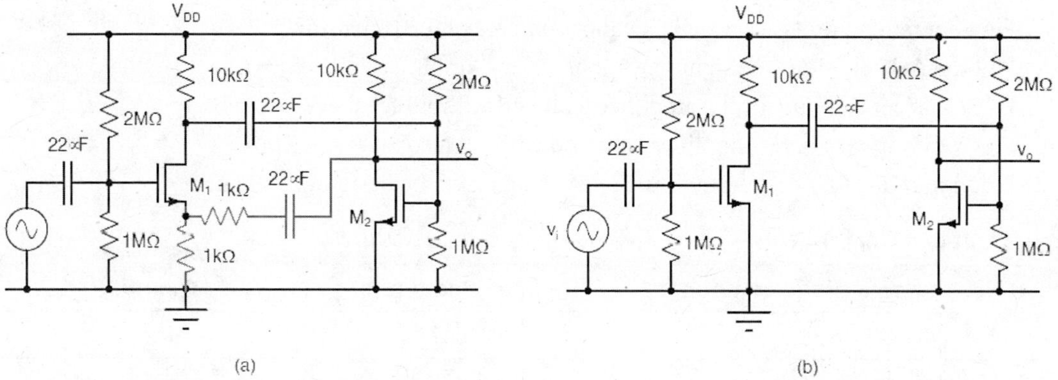

Fig. 5.25: (a) Circuit diagram of voltage series feedback amplifier where feedback part is shown in blue (b) Circuit diagram of amplifier circuit shown in part(a) with feedback element removed

1. Connect the circuit as shown in Fig.5.25(a) on breadboard.
2. Apply a sinusoidal input of 10mV amplitude and 1kHz frequency. Measure the input and output voltage using oscilloscope and calculate the small signal gain of the circuit.
3. Change the frequency from 10Hz to 100MHz in several steps to estimate the gain across frequency range.
4. Plot the AC frequency response and phase response from the observed input and output waveform on the oscilloscope.
5. Now apply a square wave input of 1kHz at the input of amplifier and plot its output response. The response at the output during input transition depicts the step response of this amplifier
6. Perform the steps 2 to 5 again for the circuit shown in Fig.5.25(b)
7. Compare these results with simulation results obtained in experiment 5.1.
8. Repeat the steps 2 to 7 for the circuit shown in Fig.5.26.

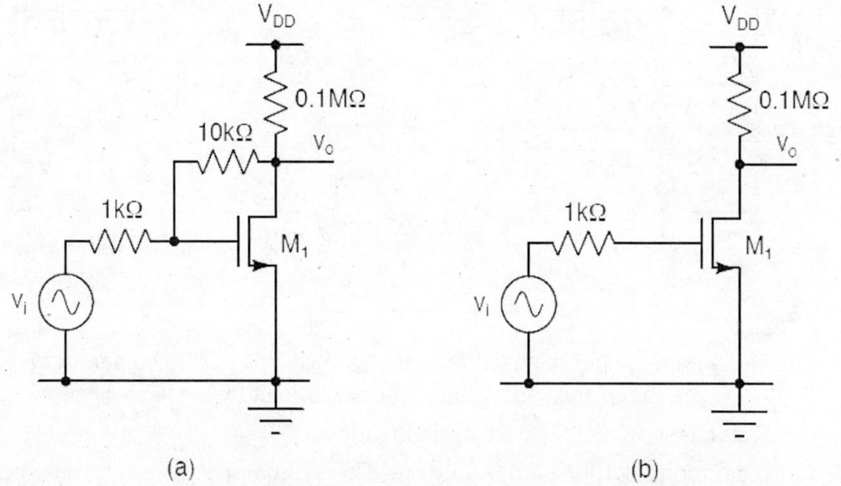

Fig. 5.26: (a) Circuit diagram of voltage shunt feedback amplifier (b) Circuit diagram of voltage series feedback amplifier with feedback element removed

Experiment-5.4: Simulation of Current Series and Current Shunt Feedback Amplifiers

Aim: Following are the aim of this simulation

1. Find operating points and small signal parameters of the transistor
2. Plot the AC magnitude and phase response of the amplifier with and without feedback
3. Plot transient response of the amplifier with and without feedback
4. Plot step response of the amplifier with and without feedback
5. Find input impedance, output impedance and gain of the designed amplifier

Equipment and Components:

S.No.	Name of the component	Specification	Quantity
1	NPN transistor	BC547	2
2	Resistors	1kΩ	1
3	Resistors	10kΩ	3
4	Resistors	2MΩ	2
5	Resistors	1MΩ	2
6	Capacitors	22μF	3
7	Breadboard		1
8	Signal source		1
9	Power supply		1

Procedure:

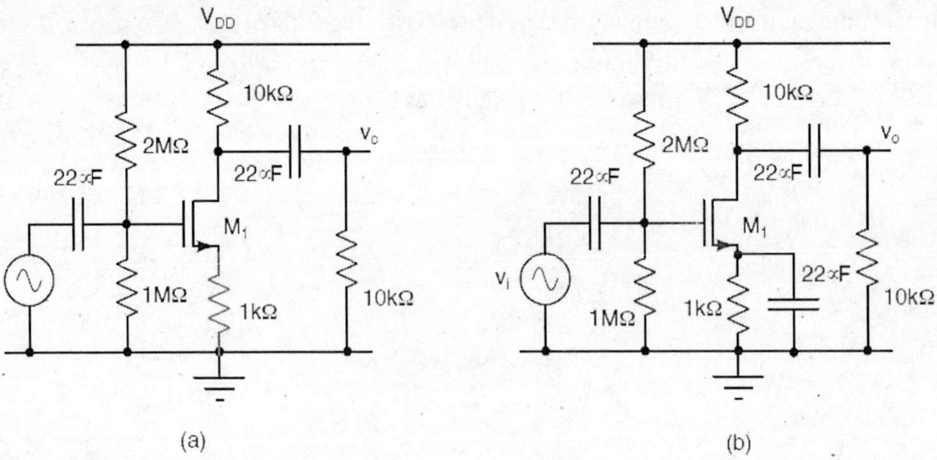

Fig. 5.27: (a) Circuit diagram of current series feedback amplifier (b) Circuit diagram of current series feedback amplifier with feedback element bypassed

1. Connect the circuit as shown in Fig.5.27(a) on breadboard.
2. Apply a sinusoidal input of 10mV amplitude and 1kHz frequency. Measure the input and output voltage using oscilloscope and calculate the small signal gain of the circuit.

3. Change the frequency from 10Hz to 100MHz in several steps to estimate the gain across frequency range.
4. Plot the AC frequency response and phase response from the observed input and output waveform on the oscilloscope.
5. Now apply a square wave input of 1kHz at the input of amplifier and plot its output response. The response at the output during input transition depicts the step response of this amplifier
6. Perform the steps 2 to 5 again for the circuit shown in Fig.5.27(b)
7. Compare these results with simulation results obtained in experiment 5.1.
8. Repeat the steps 2 to 7 for the circuit shown in Fig.5.28.

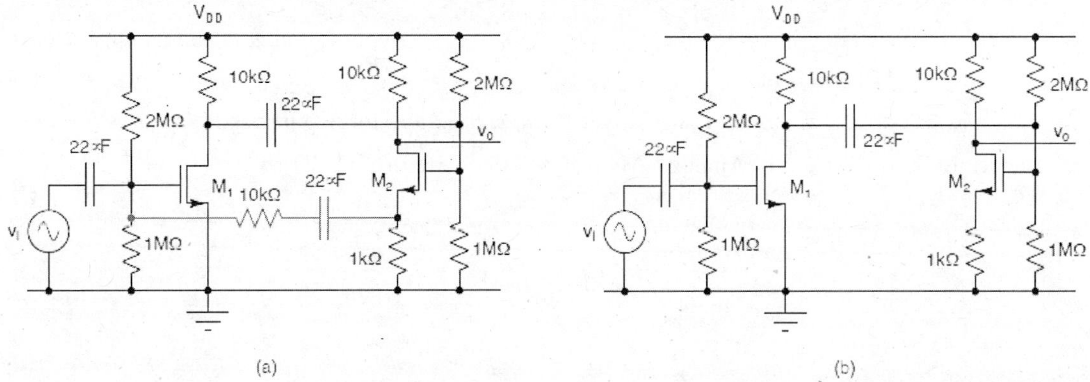

Fig. 5.28: (a) Circuit diagram of current shunt feedback amplifier (b) Circuit diagram of amplifier in part(a) with feedback element removed

KNOW MORE

The advancement in CMOS process technology has led to smaller dimension of MOS devices resulting in poor output impedance and thus limited gain offered by amplifiers designed in these processes. This has motivated designers to look for alternative methods to process signals and achieve gain. One such processing is time domain processing where instead of processing the signal in voltage domain, it is first converted into a proportional time domain signal. This time domain signal is then processed using time domain circuitry such as VCOs or delay cells. As the time resolution of advanced CMOS process nodes is becoming finer and hence time domain processing is advantageous in these advance process nodes.

Dynamic QR Code for Further Reading

REFERENCES AND SUGGESTED READINGS

1. Tsividis, Y. P. "Design considerations in single-channel MOS analog integrated circuits-A tutorial." *IEEE Journal of Solid-State Circuits* 13.3 (1978): 383-391.
2. Johns, D. A., & Martin, K. (2008). *Analog integrated circuit design*. John Wiley & Sons.
3. Tsividis, Y. (2002). *Mixed analog-digital VLSI devices and technology*. World Scientific.
4. Allen, P. E., Dobkin, R., & Holberg, D. R. (2011). *CMOS analog circuit design*. Elsevier.
5. Jespers, P. G., & Murmann, B. (2017). *Systematic design of analog CMOS circuits*. Cambridge University Press.
6. Gray, P. R., Hurst, P. J., Lewis, S. H., & Meyer, R. G. (2024). *Analysis and design of analog integrated circuits*. John Wiley & Sons.
7. Baker, R. Jacob. *CMOS: circuit design, layout, and simulation*. John Wiley & Sons, 2019.
8. Razavi, Behzad. "Design of Analog CMOS Integrated Circuits." (2000).

REFERENCES FOR FURTHER LEARNING

1. W. Shockley, "Electrons and Holes in Semiconductors" Princeton N.J.: Van Nostrand. 1950 pp.111-114.
2. Horowitz, Paul; Winfield Hill (1989). *The Art of Electronics, 2nd Ed.* London: Cambridge University Press. p. 44.
3. Acosta, Orlando N. "Zener diode-a protecting device against voltage transients." *IEEE Transactions on Industry and General Applications* 4 (1969): 481-488.
4. Shockley, William. "The Theory of p-n Junctions in Semiconductors and p-n Junction Transistors." *Bell system technical journal* 28, no. 3 (1949): 435-489.
5. Tsividis, Y. P. "Design considerations in single-channel MOS analog integrated circuits-A tutorial." *IEEE Journal of Solid-State Circuits* 13.3 (1978): 383-391.
6. Bardeen, John, and Walter Houser Brattain. "Physical principles involved in transistor action." *Physical Review* 75, no. 8 (1949): 1208.
7. Riordan, Michael, Lillian Hoddeson, and Conyers Herring. "The invention of the transistor." *Reviews of Modern Physics* 71, no. 2 (1999): S336.
8. Shockley, William. "Transistor physics." *American Scientist* 42, no. 1 (1954): 41-72.
9. Nelson, Richard. "The link between science and invention: The case of the transistor." In *The rate and direction of inventive activity: Economic and social factors*, pp. 549-584. Princeton University Press, 1962.
10. Shockley, William. "Transistor technology evokes new physics." *Nobel lecture* 11 (1956).
11. Shockley, William. "The path to the conception of the junction transistor." *IEEE Transactions on Electron Devices* 23, no. 7 (1976): 597-620.
12. Shockley, William. "The path to the conception of the junction transistor." *IEEE Transactions on Electron Devices* 31, no. 11 (1984): 1523-1546.
13. Sedra, A. S. & Smith, K. C. (2004). *Microelectronic Circuits* (5th ed.). Oxford University Press. p. 250, Eq. 4.14
14. Johns, D. A., & Martin, K. (2008). *Analog integrated circuit design*. John Wiley & Sons.
15. Tsividis, Y. (2002). *Mixed analog-digital VLSI devices and technology*. World Scientific.
16. Allen, P. E., Dobkin, R., & Holberg, D. R. (2011). *CMOS analog circuit design*. Elsevier.
17. Jespers, P. G., & Murmann, B. (2017). *Systematic design of analog CMOS circuits*. Cambridge University Press.
18. Gray, P. R., Hurst, P. J., Lewis, S. H., & Meyer, R. G. (2024). *Analysis and design of analog integrated circuits*. John Wiley & Sons.
19. Baker, R. Jacob. *CMOS: circuit design, layout, and simulation*. John Wiley & Sons, 2019.
20. Razavi, Behzad. "Design of Analog CMOS Integrated Circuits." (2000).

❑❑

CO AND PO ATTAINMENT TABLE

Course outcomes (COs) for this course can be mapped with the programme outcomes (POs) after the completion of the course and a correlation can be made for the attainment of POs to analyze the gap. After proper analysis of the gap in the attainment of POs necessary measures can be taken to overcome the gaps.

Table for CO and PO attainment

Course Outcomes	Attainment of Programme Outcomes (1- Weak Correlation; 2- Medium Correlation; 3- Strong Correlation)						
	PO-1	PO-2	PO-3	PO-4	PO-5	PO-6	PO-7
CO-1	-	-	-	-	-	-	-
CO-2	-	-	-	-	-	-	-
CO-3	-	-	-	-	-	-	-
CO-4	-	-	-	-	-	-	-
CO-5	-	-	-	-	-	-	-

The data filled in the above table can be used for gap analysis.

INDEX

❑❑